AF745553

Pseudomonas Aeruginosa - An Armory Within

Edited by Dinesh Sriramulu

Published in London, United Kingdom

IntechOpen

Supporting open minds since 2005

Pseudomonas Aeruginosa - An Armory Within
http://dx.doi.org/10.5772/intechopen.77869
Edited by Dinesh Sriramulu

Contributors
Salah Abdelbary, Marion Skalweit, Sandra Grumelli, Mauricio Corredor, Diana Carolina Castaño, Jaison Hernán Cuartas, Pablo Valderrama-Carmona, Travis Walker, Uranbileg Daalkhaijav, Angela Dunham, Orji Frank Anayo, Onyemali Chidi Peter, Ukaegbu Gray Nneji, Obinna Ajunwa, Chika Ezeanyanaso, Lawal Oluwabusola Mistura, Dinesh Sriramulu

© The Editor(s) and the Author(s) 2019
The rights of the editor(s) and the author(s) have been asserted in accordance with the Copyright, Designs and Patents Act 1988. All rights to the book as a whole are reserved by INTECHOPEN LIMITED. The book as a whole (compilation) cannot be reproduced, distributed or used for commercial or non-commercial purposes without INTECHOPEN LIMITED's written permission. Enquiries concerning the use of the book should be directed to INTECHOPEN LIMITED rights and permissions department (permissions@intechopen.com).
Violations are liable to prosecution under the governing Copyright Law.

(cc) BY

Individual chapters of this publication are distributed under the terms of the Creative Commons Attribution 3.0 Unported License which permits commercial use, distribution and reproduction of the individual chapters, provided the original author(s) and source publication are appropriately acknowledged. If so indicated, certain images may not be included under the Creative Commons license. In such cases users will need to obtain permission from the license holder to reproduce the material. More details and guidelines concerning content reuse and adaptation can be found at http://www.intechopen.com/copyright-policy.html.

Notice
Statements and opinions expressed in the chapters are these of the individual contributors and not necessarily those of the editors or publisher. No responsibility is accepted for the accuracy of information contained in the published chapters. The publisher assumes no responsibility for any damage or injury to persons or property arising out of the use of any materials, instructions, methods or ideas contained in the book.

First published in London, United Kingdom, 2019 by IntechOpen
IntechOpen is the global imprint of INTECHOPEN LIMITED, registered in England and Wales, registration number: 11086078, The Shard, 25th floor, 32 London Bridge Street
London, SE19SG - United Kingdom
Printed in Croatia

British Library Cataloguing-in-Publication Data
A catalogue record for this book is available from the British Library

Additional hard and PDF copies can be obtained from orders@intechopen.com

Pseudomonas Aeruginosa - An Armory Within
Edited by Dinesh Sriramulu
p. cm.
Print ISBN 978-1-78923-815-0
Online ISBN 978-1-78923-816-7
eBook (PDF) ISBN 978-1-83962-214-4

For EU product safety concerns:
IN TECH d.o.o., Prolaz Marije Krucifikse Kozulić 3, 51000 Rijeka, Croatia, info@intechopen.com or visit our website at intechopen.com.

We are IntechOpen, the world's leading publisher of Open Access books

Built by scientists, for scientists

4,300+
Open access books available

116,000+
International authors and editors

130M+
Downloads

151
Countries delivered to

Our authors are among the
Top 1%
most cited scientists

12.2%
Contributors from top 500 universities

WEB OF SCIENCE™

Selection of our books indexed in the Book Citation Index
in Web of Science™ Core Collection (BKCI)

Interested in publishing with us?
Contact book.department@intechopen.com

Numbers displayed above are based on latest data collected.
For more information visit www.intechopen.com

Meet the editor

Dinesh Sriramulu has a doctorate degree in Medical Microbiology from the Technical University of Braunschweig, Germany. He began his research career at the Helmholtz Centre for Infection Research, Braunschweig, Germany, in collaboration with Karolinska Institutet, Stockholm, Sweden. His area of expertise is the adaptation of bacteria towards diverse niches, ranging from the human lung to cattle rumen. In addition, he has expertise in tumor microenvironments associated with esophageal and breast cancers. Dr. Sriramulu has conducted research work at various institutions worldwide, including the University College Cork, Ireland; University of Medicine and Dentistry, New Jersey, USA; Food and Drug Administration, Rockville, USA; University of Southern California, Los Angeles, USA; University of Trento, Italy; and University of Cape Town, South Africa. He has published his research findings in various international peer-reviewed journals and presented his work at international conferences. He serves as editorial board member, peer reviewer, and expert referee for scientific journals and research funding agencies. In recent years, he has embraced the law profession and has been practicing law before the Hon'ble Madras High Court, India.

Contents

Preface

Pseudomonas aeruginosa is one bacterium among many that displays fascinating mechanisms of adaptation and evolution. With the mosaic nature of its genome, *P. aeruginosa* is able to thrive on both biotic and abiotic niches, including drinking water, invoking adaptive phenotypes that are well suited for specific microenvironments. The presence of acquired genetic elements, large and small, distributed throughout the genome requires this organism to be able to survive using a wide range of nutrients. The interdisciplinary approach to understanding various strains and isolates of *P. aeruginosa* has unraveled interesting strategies that this organism employs to combat the host defense mechanisms and thrive in unusual niches. The association of *P. aeruginosa* with cystic fibrosis (CF) is by virtue of the bacteria's colonization and formation of biofilms. The CF condition is characterized by abnormal transport of chloride and sodium across the epithelium that leads to thickening of secretions, especially in the lungs, pancreas, liver, and intestine. The complex nature of this disease, involving dysfunction of multiple organs and subsequent secondary infections by microbes, especially *P. aeruginosa*, is the basis for mortality in the CF population. This book is a compilation of chapters that highlight different facets of the capabilities and lifestyle of *P. aeruginosa*.

I would like to thank IntechOpen for appointing me editor of this book and for providing me the opportunity to contribute to the scientific community. I am also grateful to the authors of the chapters for their valuable contributions. Finally, I wish to thank the Helmholtz Centre for Infection Research, Braunschweig, Germany, where I began my research career.

Dr. Dinesh Sriramulu
Shres Consultancy,
India

Chapter 1

Introductory Chapter: *Pseudomonas aeruginosa* - Toward Omnipresence

Dinesh Sriramulu

1. Master in evolving

Antibiotics are extensively used worldwide for treating predominantly gram-negative bacterial infections and also for treating certain gram-positive infections. While the precise mechanism of their bactericidal action is yet to be unraveled, aminoglycosides, for example, act by binding to the RNA component of ribosomes, leading to both mistranslation and ultimate inhibition of protein synthesis. The widespread use of other major classes of antibiotics has resulted in the emergence of resistant bacteria by expediting the course of its evolution [1, 2]. The emergence of resistance to antibiotics is of special concern in the treatment of infections, particularly of systemic nature, by gram-negative organisms narrowing down the options for antibiotic alternatives. The resistance mechanisms displayed by the bacteria can be classified into the following: (a) reduced uptake, (b) increased efflux, (c) enzymatic modification of drug, and (d) drug target modification. Whereas resistance to streptomycin, the first widely used aminoglycoside, is predominantly through mutations in drug targets (mostly in the ribosomal protein rpsL and also in rRNA), resistance to other aminoglycosides appears to utilize a variety of mechanisms. The question arises, whether antibiotic action facilitates the emergence of resistant mutants. For certain other classes of antibiotics that induce the bacterial SOS response either by direct DNA damage (e.g., ciprofloxacin) or through indirect means (e.g., ampicillin), it has been shown that the action of the antibiotic itself plays a significant role in the emergence of mutations that confer resistance. One such mechanism, mistranslation due to defects in the translation apparatus, can promote hypermutagenesis in a phenomenon called translational stress-induced mutagenesis (TSM) raising the possibility that aminoglycoside exposure, by promoting mistranslation, could also elevate mutagenesis. According to the current understanding, TSM is mediated by a low-level mistranslational corruption of the replicative DNA polymerase leading to episodic hypermutagenesis. Exposure of wild-type bacterial cells to sublethal concentrations of an antibiotic increases mutagenic translesion DNA synthesis in vivo, and exposure of certain mutants also increases spontaneous mutagenesis. Exposure of wild-type *Pseudomonas aeruginosa* PAO1 cells to sublethal concentrations of tobramycin and amikacin, two aminoglycoside antibiotics commonly used to treat *P. aeruginosa* infections, can elevate spontaneous mutagenesis leading to complications in treating cystic fibrosis patients [3].

2. Master in dominating

Cystic fibrosis (CF) is an autosomal recessive genetic condition among Caucasians, with an incidence rate of 1 in 2500 live births. The morbidity and

IntechOpen

mortality associated with this disease condition are due to thickened lung secretions and subsequent creation of hypoxia and secondary infections predominantly by opportunistic pathogens. Bacteria such as *Pseudomonas aeruginosa*, *Staphylococcus aureus*, and *Burkholderia cepacia* complex have been in the limelight as the pathogens that affect CF patients with progression of lung disease ultimately leading to mortality. Interestingly, recent developments in high-throughput genomic techniques revealed the presence of several other bacterial species, which were hardly identified using conventional microbiological techniques. Enteric bacteria, such as *Prevotella*, *Bacteroides*, *Fusobacterium*, *Mycoplasma*, *Ralstonia*, *Veillonella*, etc., which do not normally appear in the laboratory cultures were identified and were found to have an impact on the CF microbiome [4]. This fluctuation in the CF microbiome may be due to transition of atypical species toward chronic mode of infection through formation of biofilms, dormancy, small colony variants, etc. The immunocompromised nature of CF patients predisposes them to a variety of infections, thereby increasing the need for antibiotics, alone or in combination, on a daily basis, at milligram levels. Such a continuous antibiotic pressure drives evolution of lung pathogens through the downregulation of acute-mode virulence factors in order to avoid unnecessary energy loss and expression of regulons associated with chronic mode of infection/colonization. Though the CF microbiome has been shown to consist of several species of bacteria, *P. aeruginosa* becomes the predominating one during the course of chronic colonization in the CF lung, thereby increasing its significance when considering appropriate treatment strategy [5].

Apart from the abovementioned bacterial species in the CF microbiome, mycobacteria, in general, are widespread organisms except tuberculosis (*Mycobacterium tuberculosis*) and leprosy (*M. leprae*) pathogens which are obligate parasites always in need of a host. These bacteria are often involved in asymptomatic infections, are highly fastidious organisms showing resistance to antibiotics, and are able to survive for long periods in acids, alkali, detergents, etc. Non-tuberculous mycobacteria constitute all the other mycobacteria gaining importance in respiratory infections including the one resembling tuberculosis. Practically, overgrowth of pseudomonads and other predominant bacterial species in the lung makes it difficult to understand the existence of atypical bacteria in the case of CF lung infections. The inherent slow growth rate of mycobacteria adds to failures in preliminary detection of these bacteria. Once identified it requires a prolonged treatment regime for several months with combined antibiotics, which add stress to the CF lung environment, thereby resulting in a progressive deterioration of lung function with consequent emergence of resistant pathogens. Dominance by *P. aeruginosa* or few known predominant bacterial species in the CF lung is clinically beneficial in the sense that these outnumbered species may offer protection against more pathogenic species such as mycobacteria [6].

3. A friend or a foe?

Normally, tuberculosis is rare among CF patients, and it was found to complicate the CF disease condition. It is also interesting to know that the CF disease condition will not support growth of TB mycobacteria (*Mycobacterium tuberculosis*) and the risk of TB in these patients is high in areas with high prevalence. In addition, other chronic illnesses such as poorly controlled diabetes were considered as an additional risk factor among CF patients.

Among non-typical mycobacteria (NTM), *M. abscessus* is considered the most clinically virulent species. Isolation of NTM is common in CF patients before lung transplantation as revealed by data from a large US center. However, reports mention high variability in infection rates predominantly with single species and rarely by two mycobacteria due to several factors from diversity in methodology, number of patients involved in the studies, geographical and racial differences, and the age factor, with some ambiguity in the case of gender basis. Adolescents and young adults (10–25 years age) are prone to NTM infections with rapidly growing strains infecting patients of all age groups. The slow-growing species *M. avium*, *M. intracellulare*, and other genetically related species are prevalent in North America, whereas the rapidly growing *M. abscessus* prevails in Western Europe and Israel. Infrequent prevalence was reported for *M. fortuitum*, *M. gordonae*, *M. kansasii*, *M. simiae*, *M. peregrinum*, and *M. malmoense* [7]. However, survival of CF patients infected with NTM before transplantation is reported to be similar to that of patients without NTM infection. Overall, the predominance of *P. aeruginosa* could help keep a check on infections by other pathogenic bacteria [8].

4. The concern

Huge genetic repertoire and mosaic genome structure of *P. aeruginosa* make it a versatile opportunistic pathogen in nosocomial settings, particularly in conditions involving burns and wounds, meningitis, endocarditis, and microbial keratitis. Interestingly, *P. aeruginosa* displays a common phenotype in the CF lung irrespective of the genetic content, which includes mucoidy, lipopolysaccharide modifications, lack of flagella and pili, upregulated antibiotic efflux, etc. New forms of emerging resistance in bacteria spread rapidly by intra- and interspecies acquisition of genetic content from the environment where community biofilms are common. In addition to being a threat to public health, highest resistance rates correlate with highest per capita antibiotic consumption of a nation.

Author details

Dinesh Sriramulu
Shres Consultany, India

*Address all correspondence to: d.sriramulu@gmail.com

IntechOpen

© 2019 The Author(s). Licensee IntechOpen. This chapter is distributed under the terms of the Creative Commons Attribution License (http://creativecommons.org/licenses/by/3.0), which permits unrestricted use, distribution, and reproduction in any medium, provided the original work is properly cited. CC BY

References

[1] Winstanley C, O'Brien S, Brockhurst MA. Pseudomonas aeruginosa evolutionary adaptation and diversification in cystic fibrosis chronic lung infections. Trends in Microbiology. 2016;**24**(5):327-337

[2] Moradali MF, Ghods S, Rehm BH. Pseudomonas aeruginosa lifestyle: A paradigm for adaptation, survival, and persistence. Frontiers in Cellular and Infection Microbiology. 2017;**7**:39

[3] Stefani S, Campana S, Cariani L, et al. Relevance of multidrug-resistant Pseudomonas aeruginosa infections in cystic fibrosis. International Journal of Medical Microbiology. 2017;**307**(6):353-362

[4] Cox MJ, Allgaier M, Taylor B, et al. Airway microbiota and pathogen abundance in age-stratified cystic fibrosis patients. PLoS One. 2010;**5**:e11044

[5] Rossi GA, Morelli P, Galietta LJ, Colin AA. Airway microenvironment alterations and pathogen growth in cystic fibrosis. Pediatric Pulmonology. 2019;**54**(4):497-506

[6] Lobo LJ, Noone PG. Respiratory infections in patients with cystic fibrosis undergoing lung transplantation. Lancet Respiratory Medicine. 2014;**2**:73-82

[7] Chalermskulrat W, Sood N, Neuringer IP, Hecker TM, Chang L, Rivera MP, et al. Non-tuberculous mycobacteria in end stage cystic fibrosis: implications for lung transplantation. Thorax. 2006;**61**:507-513

[8] Jones AM. Which pathogens should we worry about? Paediatric Respiratory Reviews. 2019 [epub ahead of print]

Chapter 2

Effects of Medium Components on the Bulk Rheology and on the Formation of Ferning Patterns for Biofilm of *Pseudomonas aeruginosa*

Uranbileg Daalkhaijav, Angela L. Dunham and Travis W. Walker

Abstract

Pseudomonas aeruginosa virulence and success within a broad range of hosts are largely due to the strength of its biofilms. The rheology of biofilm of *P. aeruginosa* was measured to investigate the bacterial response to nutritional conditions (medium that was modified with glycerol, glucose, sucrose, sodium chloride, and silver nitrate). The elastic modulus and the yield stress of the biofilm of *P. aeruginosa* increased in response to increases in glycerol, glucose, and sodium chloride. Alternatively, silver nitrate and glycerol inhibited biofilm formation at concentrations that were greater than 0.1 mM and 10 v/v%, respectively. Ferning patterns form as a result of diffusion-limited desiccation of the salt-macromolecule solution. Ferning coverage of about 50% and an orthogonal ferning pattern with 3° of branching were found for most of the biofilm samples. The complexity increased with modifications that caused strengthening of the biofilm, while the coverage and complexity dropped to zero when no biofilm growth was observed. The birefringent bundles of liquid crystals in the biofilm gained a new level of complexity and order within the ferning pattern that correlates with the biofilm robustness as characterized by its rheology, and these properties are heavily influenced by the nutritional environment of *P. aeruginosa*.

Keywords: biofilms, biorheology, ferning, birefringence

1. Introduction

A vast majority of microorganisms in the world exist within biofilms, which are weak hydrogels that often form at various interfaces [1]. The biofilms consist of up to 98% water, and they are typically composed of polymicrobial aggregates that are encased in extrapolymeric substances (EPS) [2–5]. Besides acting as a protective barrier, the EPS, which is made of DNA, proteins, and polysaccharides, aid in adhesion and water retention [4]. *Pseudomonas aeruginosa* is a well-known opportunistic human pathogen that is a common cause of hospital-acquired infections in burn wounds and eyes [6–8], and it is known to create persistent infection in cystic fibrosis (CF) patients [8–15], having resistance to many classes of antibiotics

[16–18]. PAO1, a medically-relevant strain of *P. aeruginosa* that is used in this study, acts as the model for biofilm-forming bacteria. To grow, bacterial cultures need water, a source of carbon, a source of nitrogen, and trace amounts of salts. The lysogeny broth (LB) is a complex and non-selective medium; many different types of bacteria can grow on non-selective medium. Lysogeny broth was formulated by Giusseppe Bertani in 1951 to optimize *Shigella* growth, but it has since become the standard for growing many bacterial cultures [19]. Lysogeny broth is composed of: 1% tryptone (source of amino acids); 0.5% yeast extract (source of vitamins, amino acids, nitrogen, and carbon); [20] and 1% NaCl (provides osmotic balance) [21]. Yeast extract is made from baker's yeast (*Saccharomyces cerevisiae*) grown to a high concentration and then exposed to high temperature or osmotic shock, killing the yeast and starting autolysis of the cells through the yeast's own enzymes [20, 22, 23]. The resulting extract solution is further filtered and spray-dried into a powder [20]. Proteins make up the most significant component of the powdered yeast extract at 62.5–73.8 wt% [20]. The average molecular weight of the yeast extract is 438 Da with 59.1% of the total under 300 Da [23]. Using additions of glycerol, glucose, sucrose, sodium chloride (NaCl), and silver nitrate ($AgNO_3$), this paper investigates various modifications of the LB medium for their effects on the biofilm.

Both the biofilm's structure and the cell-to-cell communication mechanism of the bacteria, known as quorum sensing (QS), are affected by their environment and the medium composition [24]. Quorum sensing controls additional properties that influence biofilm structures of bacteria, such as the production of extracellular DNA, proteins, mucus, and lipids [24–26]. When the growth environment becomes more viscous through the addition of glycerol, strains of *Pseudomonas* produced high-molecular-weight EPS and developed more robust biofilms [27]. The nutritional condition, such as the carbon source, influences the QS-associated swarming motility of *P. aeruginosa* [25]. While glucose supplementation limits bacterial motility, producing scattered, mushroom-like microcolonies, increasing the concentration of glucose from 0 to 2.7% caused an increase in the overall formation of biofilm [24, 25, 28, 29].

High osmolarity had a detrimental effect on biofilm of *P. fluorescens*, at roughly 0.4 Osm L^{-1} of either NaCl or sucrose, and the formation of biofilm decreased by four-fold as compared to lower concentrations of each component [30]. Similarly, mutant strains of *P. aeruginosa* that are found in CF patients transition from a non-mucoid to an alginate-overproducing state under osmotic stress that is induced by concentrations of 0.2–0.5 M NaCl (~1.2–3%) or 10% sucrose [31]. Silver has broad-spectrum antimicrobial effects on gram-negative bacteria that are well-documented [32, 33]. For instance, for concentrations of silver sulfadiazine that are lower than 0.16 $\mu g\ mL^{-1}$, planktonic growth of *P. aeruginosa* was unchanged; however, at or above this threshold amount, the concentration of the planktonic bacteria was reduced by five orders of magnitude [34]. Silver sulfadiazine was even effective against mature biofilms above a threshold dose of 1 $\mu g\ mL^{-1}$, and at concentrations of 10 $\mu g\ mL^{-1}$, it can completely eradicate a pre-established biofilm of *P. aeruginosa* [34].

The following sections of this study cover three different methods of characterizing biofilms: (i) rheology to quantify the impact of the modified medium on the mechanical strength of the biofilm; (ii) ferning to characterize the mass transport of the salts through the polymer matrix of the biofilm during desiccation; and (iii) birefringence to observe self-assembly behavior of the solute in the biofilm.

1.1 Rheology of biofilms

The study of flow and deformation of matter (rheology) enables characterization of its structure and mechanical properties. Rheology is an especially valuable tool for understanding a vast range of "soft matter" that falls between liquid and solid phases [35]. Soft matter can be divided into four classes: (1) polymers, a long repeating chain of monomers which for biological samples include proteins, DNA, and cellulose; (2) colloids, a large category of materials that describe a suspension of one material into another medium such as aerosols, foams, emulsions, suspensions, and pastes; (3) amphiphiles, molecules with dual characteristics where one end of the molecule likes the solvent (hydrophilic), while the other end does not (hydrophobic) include surfactants that are amphiphiles at the air-water interface; and (4) liquid crystals, rod or disk shaped molecules that self-assemble to form orientation order but not positional order, resulting in an anisotropic fluid [35, 36].

Rheological techniques can characterize the strength and behavior of clinically relevant biological fluids such as mucus, blood plasma, and bacterial biofilm. More importantly, we can also use rheological measurements to drive the treatment of the biofluids toward a favorable clinical outcome. A rheological testing can quantify the viscous and elastic properties of a material. Two main modes of testing exist on a rotational rheometer: (1) steady-shear testing mode (**Figure 1a–d**), where the material is sheared between a stationary bottom plate and rotating top plate at a given stress or strain; and (2) the oscillation mode (**Figure 1e–g**), where the top plate oscillates back and forth at a set frequency and amplitude.

From the shear-flow sweep test, the stress (τ) versus shear rate ($\dot{\gamma}$) curve gives important information on the flow properties of the material (**Figure 1b**). The flow behavior can be modeled by Herschel-Bulkley function for yield stress fluids (Eq. (1): τ_y, yield stress; c [Pa s], flow coefficient or consistency index; p, power-law index). Yield stress materials have a minimum stress that must be surpassed before the material starts flowing. Different types of material flow behavior are described in **Figure 1b–d**.

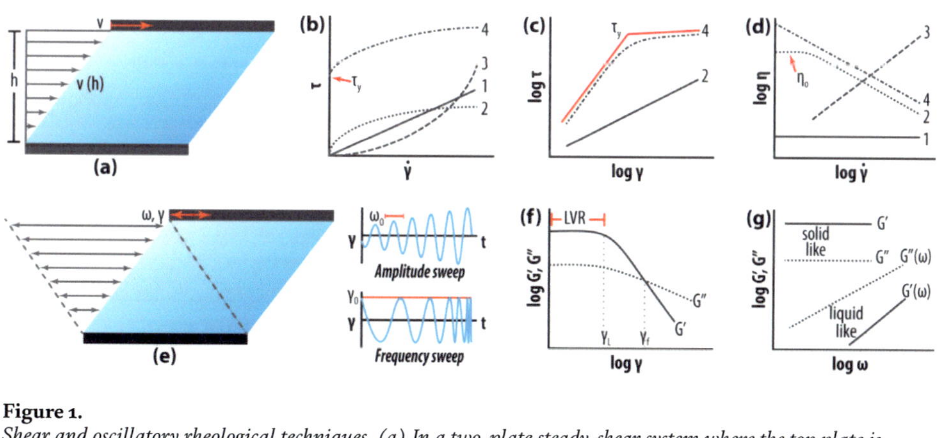

Figure 1.
Shear and oscillatory rheological techniques. (a) In a two-plate steady-shear system where the top plate is moving, the velocity (v) of the fluid is dependent on the gap height (h). (b) Samples can exhibit several different flow behaviors of stress versus strain rate, including (1) ideally viscous Newtonian fluid; (2) shear-thinning fluid; (3) shear-thickening fluid; and (4) yield stress fluid. Yield stress materials have a minimum stress (τ_y) that must be overcome before flow starts. (c) The stress-strain curve demonstrates a material with (2) no yield stress and (4) a material with a clear yield stress calculated using the tangent crossover point method. (d) The viscosity-shear rate plot shows (1) Newtonian fluid; (2) shear-thinning fluid with no yield stress reaching zero-shear viscosity (η_0); (3) shear-thickening fluid; and (4) shear-thinning fluid with yield stress. (e) In a two-plate system the top plate can oscillate back and forth at set amplitude or frequency for oscillatory rheology tests. The amplitude sweep test has constant frequency (ω_0) with changing strain amplitude while the frequency sweep test has constant strain amplitude (γ_0) with changing frequency. (f) During the amplitude sweep test, the strain values up to the limit of the strain where the G′ and G″ values are constant (γ_L) are called the linear viscoelastic region (LVR). The point where G″ crosses over G′ is called the flow point (γ_f) [72]. (g) In a frequency sweep, when G′ > G″ materials are said to be solid-like and when G″ > G′, materials are said to be liquid-like.

$$\tau = \tau_y + c * \dot{\gamma}^p \tag{1}$$

Two main types of oscillatory testing exist (**Figure 1e–g**). An amplitude sweep test oscillates the upper plate back and forth at a set frequency (ω_0) at increasing strains (**Figure 1e**). On the modulus (G′—elastic; G″—viscous) versus strain plot (**Figure 1f**), the plateau is the linear viscoelastic region (LVR), and the strain limit of the region is γ_L. A frequency sweep test oscillates at a set amplitude (γ_0), which has been determined previously from the amplitude sweep test to be within the LVR, at increasing frequency (**Figure 1e**). The frequency sweep describes how the material acts when the material is stressed for different periods. For example, when Silly Putty is stressed quickly by throwing it on the floor, it bounces back, acting like a rigid solid. However, when the Silly Putty sits at rest and experiences low stress over a long period of time, it spreads out, acting like a viscous fluid. The frequency sweep (**Figure 1g**) reveals if the material is solid-like ($G' > G''$) or liquid-like ($G'' > G'$) and if the behavior is frequency dependent ($G'(\omega)$, $G''(\omega)$) or independent. Stable gels and suspensions are typically solid-like and frequency-independent, so these types of materials are called "gel-like."

Previous studies on biofilm rheology using various techniques of rheological measurement have found the elastic modulus (G′) to range in order of magnitude from 10^{-2} to 10^4 Pa for bulk biofilms at solid-liquid interfaces using plate-on-plate methods, while the values of yield stress (τ_y) range in order of magnitude from 10^{-1} to 10^5 Pa [2, 37–42]. The wide-ranging values of G′ and τ_y in the literature reflect the variability in the compositions of the biofilms, diversity of growth mediums, variability of growth conditions, and most importantly, natural variability of response of the microorganisms, even to the same medium and growth conditions. This chapter uses the techniques established in our previously published work on the non-destructive development and characterization of rheological properties of biofilms [43]. Using this non-destructive method, the measured values of elastic modulus and yield stress of PAO1 that were grown in standard LB medium were both between 0.1 and 10 Pa [43].

1.2 Crystallization of biological materials

Biological fluids like tears, cervical mucus, and saliva are all shown to self-assemble into fractal-like patterns of crystallization when they are dried [44, 45]. A fractal is a structure that is made of smaller parts that resemble the bigger parts, with a high degree of organization and self-similarity. This structure can be characterized with a specific fractal dimension [44]. Fractal dimension is a measure of complexity of the fractal pattern [46]. Random nucleations of salts initiate the process of crystallization, where its growth is limited by the diffusion of salt through the polymer matrix (proteins or macromolecules) [47]. Therefore, the combined effects of ionic strength, osmolarity, and the size and concentration of macromolecules control the behavior of crystallization, where too little or too much of one factor can dramatically alter the pattern of crystallization [48, 49]. A typical crystallization of biosaline proceeds in the following manner: (i) salt nucleation initiates the process of crystallization; (ii) the nucleation point grows with some symmetry into a highly-branched structure whose growth is modified by the interaction of the salt with the biological matter; (iii) the branches do not overlap or merge [47, 50, 51]. The process of crystallization of biofluids is called "arborization," "ferning," or "dendritic growth" in various literature [45]. In this paper, the general formation of salt crystals will continue to be called crystallization, while the specific crystallization of the biofluids that result in fractal patterns will be called ferns.

The ferning patterns of the dried samples of tears and saliva have been used for years as a supplementary diagnostic tool [49, 52]. The ferning patterns in saliva and in tears exhibit different morphologies; saliva produces linear ferns with branching angles of 90°, while ferns from tears have more curvature with tightly packed branches with acute angles. Ferning patterns from tears and saliva are traditionally classified in a qualitative manner according to Rolando's system as Type I to Type IV [53]. Type I has the most ferning and the highest concentration of protein, while Type IV has no ferning and the lowest amount of protein [49]. Samples of tears from healthy individuals typically exhibit robust, highly-branched ferning patterns (Type I and II), while samples from patients with eye or immune diseases show little to no ferning (Type III and IV) [48, 49, 52]. Through analysis of X-ray microscopy and scanning electron microscopy (SEM), the molecular structure of the ferns from tears is revealed to be composed of NaCl, KCl, and proteins [48]. In addition to helping detect infection, saliva ferning pattern has been shown to be useful for tracking ovulation cycle from highest fertility level during estrus to lowest fertility level during diestrus [54].

Cervical mucus is a heterogeneous hydrogel that changes over the course of an animal's reproductive cycle [44, 45, 55]. Regardless of the source, human or otherwise, high levels of estrogen are produced during ovulation or peak fertility, resulting in linear ferns with branching angles of 90°, while no ferning is found during the period of low fertility when progesterone is dominant [44, 45]. During ovulation, the cervical mucus is over 98% water with its highest level of salt while both water content and salt content drops during low fertility period [45]. The low salt content during low fertility period is the cause of the lack of ferning pattern.

SEM analysis of ferns from gelatin-NaCl mixture revealed that the backbone of the ferning pattern was a series of interlocking crystalline blocks that were 10–30 μm in size [47]. When the fractal dimension of bovine cervical mucus (BCM) that was taken during ovulation was determined using the box-counting method, it was about 1.7, characteristic of diffusion-limited growth processes [44, 46]. Box counting method is based on counting non-empty boxes making up a fractal pattern on a grid [46]. A diffusion limitation was observed with the gelatin-NaCl mix as well. The ferning pattern became much less geometric and increasingly random at higher gelatin-to-NaCl ratios, where more diffusion limitation occurs due to the crosslinking of the gelatin, and the ferning ceased at extremely high gelatin-to-NaCl ratios [47]. Furthermore, the ferning pattern developed curvatures at evaporation rates above a threshold value of 11 $\mu m\ s^{-1}$ [47].

Bacterial biofilms produce ferning patterns that are similar to gelatin and mucus samples [50, 56]. Upon evaporation of droplets of solutions of various salts with cells of *E. coli* and *Bacillus subtilis*, ferning patterns emerged where the crystallized top layer covered a base layer that consisted of bacterial cells. The structure of the *E. coli* ferns was linear with branching angles of 90°, similar to cervical mucus. Neither sterile saline solutions nor *E. coli* in pure water produced ferning, confirming the previous findings that ferning results from balanced proportions of salts and macromolecules [50, 56]. Bacteria inside the crystalline structure were effectively in a state of suspended animation that was capable of reanimation after rehydration, even a week later [50]. This crystallization was hypothesized to be a form of biomineralization [50], which occurs when biological organisms produce organo-mineral hybrids that give the organism mechanical strength and hardness. Examples of biomineralization that are found in nature include bones, teeth, shells, corals, and algal silica [57]. Previous studies on strain PAO1 of *P. aeruginosa* in flow cell reactors have shown biomineralization of calcium carbonate within the EPS of the biofilm [58]. SEM of the ferning sample of *E. coli* revealed a 3D structure that was composed of dried EPS, bacteria, and salts, with the salts concentrated in the crystalline region, consistent with the previously mentioned studies [50].

Studying the ferning pattern and complexity of biofluids or biogels gives a simple and indirect measurement of the structures within the material that guide or hinder the movement of ions that ultimately form these distinct crystallization patterns. While much of the ferning patterns seen in biofluids are linear patterns with 90° branching angles, tightly packed and curved ferning patterns can be expected to develop in environments that induce fast evaporation such as in low-viscosity fluids or environments that are highly diffusion limited such as in high macromolecule to salt ratio fluids.

1.3 Birefringence of *P. aeruginosa*

One of the techniques of self-assembly for small particles is through depletion attraction in a solvent during solvent evaporation [59, 60]. Depletion attraction is an entropic force that becomes relevant when the particles in the solvent move close enough together that their excluded volumes overlap [61]. This overlap increases the osmotic pressure in the surrounding fluid and further pushes the particles together [59]. These highly ordered or anisotropic solution is described as having a liquid crystal phase and this phase is birefringent, which means that their ordered state will split light into two beams with perpendicular polarization [36, 60, 62]. Liquid crystal phases have been observed with many different types of biopolymers such as DNA, peptides, glycopolymers, proteoglycans, viruses, collagen, cellulose, phages, and chitin [60, 61, 63]. Liquid crystals form: (i) nematic phase where the molecules form directional order but no positional order; (ii) smectic phase with positional order; or (iii) chiral phase with twisting order [60]. Of these, biopolymers most commonly have nematic phase.

P. aeruginosa that exists in a viscous or anaerobic environment is stimulated to transcribe filamentous Pf bacteriophages that are about 2 μm in length and 6 nm in diameter [64]. In *P. aeruginosa* biofilm, the filamentous phage self-assembles through depletion attraction, with the biopolymers exerting the osmotic force that bundles the phage strands. These highly ordered anisotropic regions of nematic phase liquid crystals are birefringent, possessing a large negative charge, and the anisotropy was shown to increase with the ionic strength and the molecular weight [64, 65]. Birefringence is not only a direct indicator of molecular order, but it is an indicator for *P. aeruginosa* biofilm strength, surface adhesivity, desiccation tolerance, and antibiotic resistance [62, 64]. The filamentous bacteriophages facilitate chronic infection of *P. aeruginosa* in the host by promoting a less invasive, less inflammatory but more resistant, more persistent form of *P. aeruginosa* [66]. In addition, Pf phages can bind iron to inhibit the metabolic activity of other pathogens such as *Aspergillus fumigatus* [67].

Liquid crystal methods provide the means to study the structure and behavior of filamentous bacteriophages without perturbation [65]. Moreover, liquid crystal analysis, specifically through detection of its birefringence, was used to detect analytes such as glucose, cholesterol, *E. coli*, and even viruses such as Ebola and HIV [68]. This detection method was made possible through enzymatic reaction in response to analytes within the mesophase of the normally optically isotropic lipidic cubic phases that results in the formation of strongly birefringent liquid crystal phases that are easily detected optically [68]. The exogenous and endogenous birefringence from various classes of analytes were exploited to make simple and cheap detection tool that was proposed as a new diagnostic tool that can be utilized in industry or in the field to detect biothreats [68].

1.4 Objective

Biofilm is composed of motile bacterial cells, non-motile bacterial aggregates, and mucoid hydrogels of EPS that have a heterogeneous, highly-porous microstructure, allowing diffusion of water, nutrients, waste, and electrolytes [26, 69]. A complex set of interactions between the electrolytes, solutes, bacteria, and biopolymers dictate the strength, bacterial resistance, and infection persistence of the biofilm. The objective of this study is to characterize the behavior of the bacterial culture in the presence of various environmental conditions, including a highly viscous media, nutrient-enhanced media, high osmolarity media, and antimicrobial media. The interaction of the bacterial culture with its nutrient environment is measured as a function of the strength of its biofilm through rheological analysis, while its ferning pattern characterizes the mass transport through the environment in the biofilm. Additionally, birefringence inside a biofilm provides insight into the solute interaction with the biofilm.

2. Materials and methods

2.1 Bacterial strains, media, and growth conditions

The strain of *Pseudomonas aeruginosa* that was used for the entire study was the laboratory-adapted wild-type strain, PAO1. Miller lysogeny broth (LB) was prepared from BD Difco dry powder and autoclaved. Five different types of chemical modifications were made to the lysogeny broth: (i) glycerol was added to form between 1 and 15 v/v% in LB medium; (ii) glucose was added to form concentrations between 0.5 and 4.5 w/v% in LB; (iii) sucrose was added to form concentrations between 0.5 and 4.5 w/v% in LB; (iv) NaCl added to form concentrations between 1.5 and 5 w/v% in LB; and (v) $AgNO_3$ added to form concentrations between 0.001 and 1 mM in LB. Modified LB medium in a petri dish (3.6 mL) was inoculated with an overnight culture of PAO1 (0.4 mL) and was incubated for 6 days at 37°C. Some of the dishes of modified LB medium were kept sterile and incubated along with the biofilm samples to act as a negative control.

2.2 Measuring bulk biofilm rheology

The sample rheology was measured on the Discovery Hybrid Rheometer 3 (DHR3, TA Instruments, USA) using a 40-mm stainless steel plate geometry at 25°C. The measurements took place in the following order:

a. Pre-stressed: 0.1 Pa, 2 minutes

b. Frequency sweep: $\gamma_0 = 0.1$ (biofilm), $\gamma_0 = 0.005$ (sterile LB), $\omega \in [0.01, 1]$ rad s^{-1}

c. Stress sweep: $\tau \in [0.01, 1]$ Pa (biofilm), $\tau \in [0.001, 0.1]$ Pa (sterile LB), terminating if the shear rate $\dot{\gamma} > 10$ s^{-1}.

Detailed description of the sample inoculation, the incubation, and the rheological measurement methods are located in our previous work [43].

2.3 Ferning properties of dried bacterial biofilms

The biofilm samples were dried in the incubator, forming ferning patterns that were large enough to be easily seen by the unaided eye. In this paper, the previous qualitative method for ferning characterization was converted to a quantitative method of image analysis by calculating the coverage area, the fractal dimension, and the complexity score (degree of branching) of the ferning pattern. This analysis was completed by taking photographs of the surface of the petri dish, converting the photographs to black and white image on MATLAB (Figure S4, https://ir.library.oregonstate.edu/concern/defaults/g158bp85b), and finally calculating the ferning coverage by determining the ratio between white and black pixel areas in the image. The fractal dimension was calculated using the box-counting method on MATLAB (Figure S4, https://ir.library.oregonstate.edu/concern/defaults/g158bp85b).

2.4 Dried bacterial biofilms under the microscope

Microscopic images of the biofilm in its liquid and its dried ferning state were taken using an Eclipse Ti-S inverted microscope (Nikon, Japan). The polarized images were produced with polarized filters.

3. Results and analysis

3.1 Rheological characterization of biofilms

3.1.1 Sweeps of frequency

Data from the sweeps of frequency showed the difference in the viscoelasticity of the samples of biofilm (filled square) and the samples of sterile LB medium (unfilled circle) that were incubated for 6 days (**Figure 2**). For brevity, the term "unmodified LB" will refer to the standard LB medium without chemical alterations, while "modified LB" will refer to any of the five chemical additions to standard LB medium (glycerol, glucose, sucrose, NaCl, and $AgNO_3$). The sweep of frequency of the biofilm showed frequency-independent, elastic modulus (G′) dominance over the viscous modulus (G″) for all of the samples, as expected for a weak gel (Figure S3, https://ir.library.oregonstate.edu/concern/defaults/g158bp85b). The complex modulus $\left(|G^*| = |\sqrt{G'^2 + G''^2}|\right)$ for the samples of biofilm (~0.02 Pa) were, on average, an order of magnitude larger than their counterparts of sterile LB medium (**Figure 2a**). Previous work on the sterile LB medium showed that aging and cycling of temperature can cause buildup of protein at the air-liquid interface, which can be measured by a plate geometry to an extent, though not as accurately as using a du Noüy ring. Therefore, the measurable viscoelasticity in the sterile samples can be attributed to the age of the medium at the time of use (>3 months) and to the incubation process, which further accelerated the aging. The complex modulus of the unmodified LB biofilm and the sterile unmodified LB medium were each averaged over the entire frequency range (0.1–1 rad s^{-1}) and were averaged between replicates (biofilm n = 12, sterile n = 7) to get mean values for G*. All the raw results from sweeps of frequency of the biofilm and of the sterile samples in modified LB medium (Figure S2, https://ir.library.oregonstate.edu/concern/defaults/g158bp85b) were averaged over the entire frequency range and plotted (Figure S3b–f, https://ir.library.oregonstate.edu/concern/defaults/g158bp85b). The mean values of G* for biofilms that were grown in unmodified LB (red) served as the base of comparison for biofilms that were grown in modified LB medium (**Figure 2b–f**). Based on the modulus data, bacterial biofilm appeared to be strongly affected by its nutritional environment.

3.1.2 Sweeps of stress

The values of yield stress were derived from the experiments of increasing strain, where the yield stress is the point of offset of a stress-versus-strain curve.

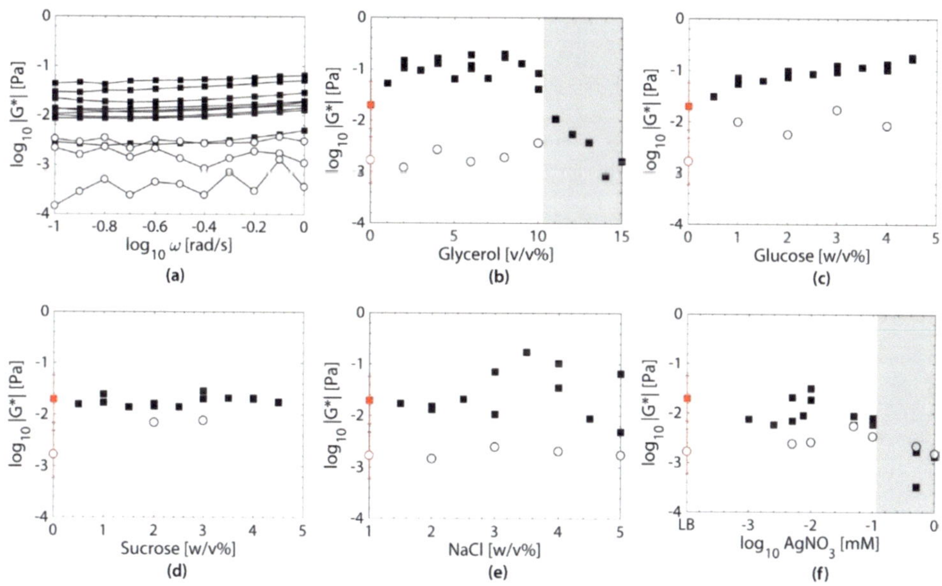

Figure 2.
Modulus |G| calculated from the sweep of frequency showing the biofilm (filled square) and sterile LB medium (unfilled circle) samples. The results of the (a) sweep of frequency from $\omega \in [0.1, 1]$ rad s^{-1} (biofilm $\gamma_o = 0.1$ and sterile LB medium $\gamma_o = 0.005$) of biofilm samples that were grown in unmodified LB medium and of sterile unmodified LB medium. The mean modulus |G*| was 0.015 Pa for biofilm and 0.0015 Pa for sterile unmodified LB medium. (b–f) The mean modulus |G*| is plotted (red) with relative errors that were calculated from standard error to ensure symmetry ($n \geq 3$). The average |G*| of biofilm that was grown in modified LB medium with the following concentrations (b) 1–15 v/v% glycerol, (c) 0.5–4.5 w/v% glucose, (d) 0.5–4.5 w/v% sucrose, (d) 1.5–5 w/v% NaCl and (e) 0.001–1 mM $AgNO_3$ are shown. The gray regions in (b) and (f) represent inhibiting concentrations that had no biofilm growth.*

The stress-versus-strain data of the unmodified LB biofilm (filled square) and sterile unmodified LB samples (unfilled circle) showed that the samples of biofilm exhibited an appreciable yield stress (**Figure 3a**). After yielding, the material stress was constant (<1 Pa) at high strain, while the samples of sterile medium demonstrated no yield stress. In fact, none of the sterile modified or unmodified LB mediums had an appreciable yield stress (Figure S2, https://ir.library.oregonstate.edu/concern/defaults/g158bp85b). The mean yield stress of the samples of unmodified LB biofilm (red) was compared with the biofilms that were grown on the modified LB medium (**Figure 3b–f**). The results of yield stress showed a similar dependence on the nutritional conditions of the biofilms as the results of the modulus from the sweeps of frequency.

3.1.3 Modification with glycerol

With the addition of glycerol (0–15 v/v%), the complex modulus of the biofilm increased by almost an order of magnitude between 0 and 2% glycerol, remained constant between 2 and 10%, and experienced a dramatic drop in modulus for concentrations greater than 10% to modulus values that are comparable to sterile LB (**Figure 2b**). The yield stress of the biofilm showed similar trends, increasing by one order of magnitude with glycerol from 0 to 10% until concentrations of glycerol that were greater than 10% impeded the growth of biofilm, also resulting in no yield stress (**Figure 3b**). The addition of glycerol increased the viscosity of the medium as well as inducing high osmolarity (1.4 Osm L^{-1} at 10%), promoting stronger biofilm. Other studies with glycerol-supplemented medium saw an increase in the production of EPS by biofilm, consistent with the present study [27, 70]. The glycerol can trigger

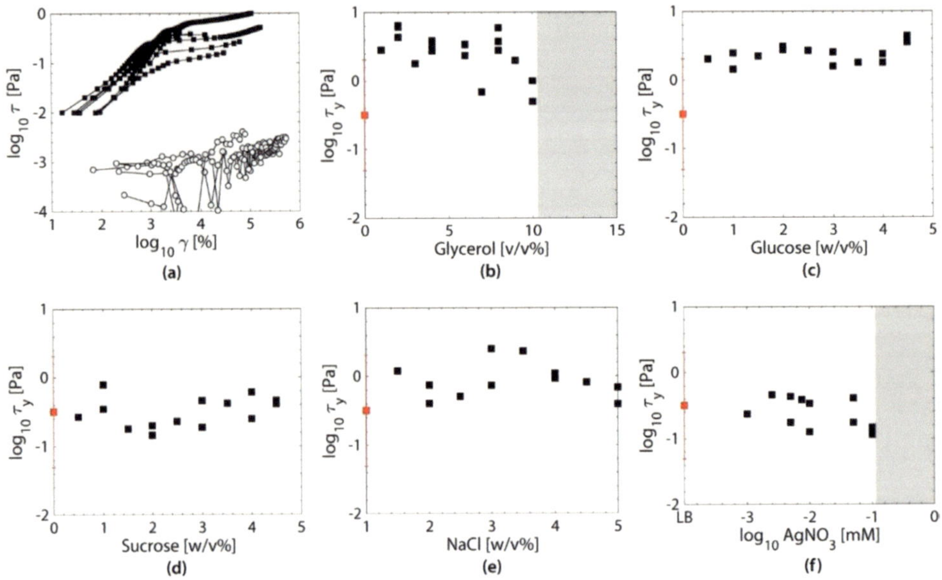

Figure 3.
(a) The stress versus strain data of biofilm that was grown in unmodified LB medium (filled square) and of samples of sterile unmodified LB medium (unfilled circle). (b–f) For the mean yield stress of biofilm that was grown in unmodified LB medium (red square), the standard error bars were converted to relative errors to ensure symmetry on the y-axis ($n \geq 7$). Plots of yield stress τ_y of the biofilms that were grown in modified LB medium at the following concentrations (b) 1–15 v/v% glycerol, (c) 0.5–4.5 w/v% glucose, (d) 0.5–4.5 w/v% sucrose, (e) 1.5–5 w/v% NaCl and (f) 0.001–1 mM $AgNO_3$ are shown. The mean τ_y was 0.32 Pa for the unmodified LB biofilms, while the LB medium did not have a yield stress. The gray regions in (b) and (f) represent inhibiting concentrations that had no biofilm growth.

pathways of production of EPS; [27] however, at high concentrations of glycerol, the diffusion-limiting environment of the highly viscous solution with high osmotic pressure (>4 Osm L^{-1} at >10%) appeared to inhibit growth. The complex modulus of the modified LB medium, on the other hand, stayed relatively constant with glycerol addition. The dramatic drop in the modulus of biofilm samples that were grown in medium that was modified with >10% glycerol corresponded with an apparent lack of biofilm in the Petri dishes, as the dishes appeared clear and yellow instead of opaque and greenish (Figure S1, https://ir.library.oregonstate.edu/concern/defaults/g158bp85b). The greenish hue in the samples is a result of the presence of pyocyanin, which is a bluish-tinted toxin that is produced by PAO1.

3.1.4 Modification with glucose

The modulus of the biofilm increased by one order of magnitude by increasing the concentration of glucose from 0 to 4.5% (**Figure 2c**), indicating that glucose was being utilized by the bacteria as an additional source of carbon which promoted growth and development of a stronger network of biofilm. The rheological results of the sterile glucose-modified LB medium did not change significantly from the unmodified LB medium. The values of yield stress followed the same trend, where the biofilm that was grown in glucose-modified LB medium had yield stresses that were an order of magnitude larger than the unmodified LB biofilm (**Figure 3c**). A previous study observed the same effect, finding that the addition of glucose up to the highest level tested, which was 2.7%, enhanced biofilm production [29]. The maximum addition of glucose (4.5%) induced osmotic pressure of 0.25 Osm L^{-1}, which did not cause inhibiting effects.

3.1.5 Modification with sucrose

Based on the rheology, sucrose did not increase biofilm production, as no change existed in the modulus (**Figure 2d**) or yield stress (**Figure 3d**) of the biofilm. In previous studies, concentrations of sucrose above 10% in medium for *P. aeruginosa* resulted in biofilm with mucoid development, while *P. fluorescens* started to experience adverse effects above 15% at which point the biofilm dramatically decreased [30, 31]. In those studies, bacterial culture reached an inhibiting level of sucrose at 15% due to osmotic pressure (0.44 Osm L^{-1}) [30]. In the present work, samples of PAO1 experienced a maximum of 0.13 Osm L^{-1} in osmotic pressure from modification with sucrose, which is well below the reported osmotic level for inhibition. *P. aeruginosa* may not be capable of utilizing sucrose, so in contrast to the simpler glucose, sucrose had little impact on the rheological properties of the biofilm.

3.1.6 Modification with sodium chloride

Unmodified LB medium already consists of sodium chloride (NaCl) at a concentration of 1%, and the modified concentration varied from 1 to 5% of NaCl. The complex modulus of biofilm remained constant for concentrations below 2.5% and increased by one order of magnitude for concentrations between 2.5 and 5%, while the modulus of the sterile modified LB medium was not affected by the concentration of NaCl (**Figure 2e**). Similarly, the yield stress increased as the concentration of NaCl was increased greater than 2.5% (**Figure 3e**). NaCl is already required for bacterial growth to provide osmotic balance, but a larger amount of salts appeared to promote stronger biofilm. The change in the biofilm could be caused by the higher salinity or osmolarity, making the environment hostile, triggering a higher level of production of alginate and other types of EPS as a countermeasure. Previous studies found that concentrations of NaCl between about 1 and 3% increased production of biofilms in *S. aureus* and *P. aeruginosa*, while concentrations of about 6% of NaCl prevented growth of biofilm in *S. aureus* [29, 31]. At concentrations of NaCl above 10%, no biofilm growth was observed, and the plate quickly crystalized to cubes of salt (Figure S6, https://ir.library.oregonstate.edu/concern/defaults/g158bp85b).

3.1.7 Modification with silver nitrate

Silver has antimicrobial properties that can inhibit bacterial growth and development of biofilm. Supplementation of silver nitrate ($AgNO_3$) to the modified LB medium has no impact on the complex modulus (**Figure 2f**) or the yield stress of the biofilm for concentrations below 0.1 mM (**Figure 3f**). Past this concentration, the modulus instantly reduced to the same level as the sterile modified LB medium, and the yield stress disappeared. Correspondingly, the plates of biofilm at the higher silver concentrations appeared clear and less viscous, resembling sterile modified LB medium (Figure S1, https://ir.library.oregonstate.edu/concern/defaults/g158bp85b). Therefore, the antimicrobial activity of the silver appeared to be strongly dependent on concentration, with little to no effect on bacterial growth at concentrations lower than the threshold (0.1 mM) and deadly at higher concentrations. These results are consistent with previous studies where the inhibitory threshold for *S. aureus* was over 0.033 mM, while the inhibitory threshold for *P. aeruginosa* was over 0.16 μg mL^{-1} (0.45 mM) [34, 71].

3.1.8 Summary of rheological characterization

The rheological parameters of elastic modulus and yield stress are useful measures of the strength of a biofilm. The complex modulus and the yield stress of the biofilms increased with the addition of glucose, which served as an additional source of carbon, but they were unaffected by addition of sucrose, which is a complex sugar that the bacteria could not utilize. The strength increased to an extent with osmolarity (glycerol and NaCl) and dramatically reduced to their sterile baseline at concentrations that were higher than the inhibitory threshold of an antimicrobial agent ($AgNO_3$). Samples with higher rheological properties correlated with a biofilm that appeared more viscous than the unmodified LB biofilm, while samples with lower modulus, and lacking a yield stress, such as high concentrations of glycerol and silver ions, appeared less viscous and free of biofilm. The values of modulus and yield stress for the samples of biofilm displayed the same medium dependent response; therefore, either measurement would be a useful metric of the strength of biofilm. Out of the five chemical modifications, three modifications increased the strength of the biofilm when compared to the unmodified LB biofilm: glycerol for concentrations up to 10%, glucose for concentrations at least up to 4.5%, and NaCl for concentrations higher than 2.5%. One of the chemicals, sucrose, had no measurable effect on the strength of the biofilm for concentrations at least up to 4.5%, while another modifier, $AgNO_3$, inhibited bacterial growth at a concentration above 0.1 mM.

3.2 Development of ferning patterns on dried plates of biofilm

After the plates were dried over a span of weeks in the incubator, they were photographed, and the photographs were converted to black and white images and cropped (Table S1, https://ir.library.oregonstate.edu/concern/defaults/g158bp85b). The patterns of crystallization on the plates were analyzed based on the ferning coverage of the plate and the complexity of the pattern (**Figure 4**). The complexity of the ferning was scored qualitatively by the degree of branching of the pattern (**Figure 4**): 0, empty plate, no ferning; 1, seed or nucleation points; 2, lines without branching; 3, orthogonal pattern with 1° of branching; 4, orthogonal pattern with 2° of branching; 5, orthogonal pattern with 3° of branching; 6, orthogonal pattern with 4° of branching; 7, orthogonal pattern with 5° of branching; and 8, dendritic pattern with branching at acute angles. The branches at acute angles were distinctly smaller, and they were irregularly branched when compared to the orthogonal ferns that were scored 3–7 on the complexity scale.

The ferning coverage was calculated quantitatively based on the percent of white pixels in the black and white images that were converted from its original photograph (**Figure 5a**). A photograph of the ferning on a plate of unmodified LB biofilm showed high ferning coverage (top photos), while the plate of sterile unmodified LB medium was noticeably absent of ferning with low calculated coverage (bottom photos). Even without biofilm growth, the sterile coverage values were not zero because the lighting and the glare of the plate surface produced some pixel artifacts. **Figure 5b–f** shows the mean coverage of the plates of unmodified LB biofilm (red filled square, n = 12) and of the plates of sterile unmodified LB medium (red unfilled circle, n = 7) plotted with the data of the modified LB medium. The left black y-axis is the ferning coverage, while the right gray y-axis is the qualitative ferning complexity score for biofilm (gray filled square) and the sterile LB medium (gray unfilled circle).

3.2.1 Modifications with glycerol

With the addition of glycerol, the ferning pattern of the samples of biofilm initially changed from a complexity score of 5 and a coverage of 47% (unmodified LB

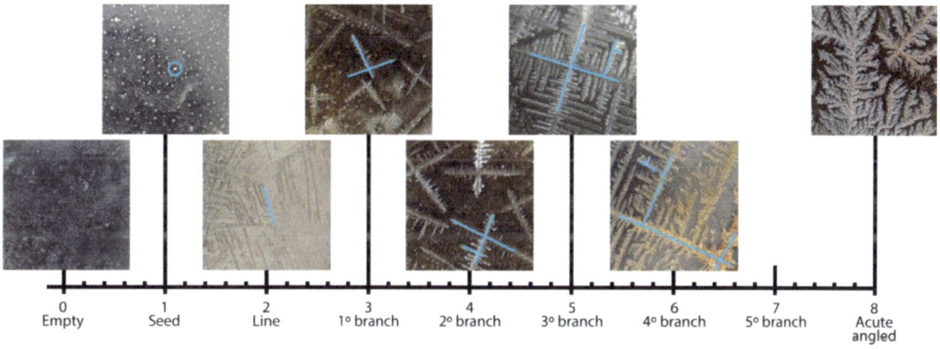

Figure 4.
A guide for the ferning complexity score of the dried biofilm ferning pattern.

biofilm) to a complexity score of 8 (**Figure 5b**). The change in ferning morphology from the orthogonal form to the acute branching form occurred at the lowest tested concentration of glycerol (**Figure 5b**: 2% plate, top left). However, as the concentration of glycerol increased, the ferning coverage dropped dramatically, reaching zero at around 8%. From both the visual inspection and the rheological measurement, samples below 10% had strong biofilm. However, a large amount of glycerol prevented the sample from completely drying, leaving the surface of the sample looking shiny and wet, causing both the ferning coverage and the complexity score to drop between 4 and 10% glycerol (**Figure 5b**: 10% plate, top right). The plates with concentrations of glycerol above 10% never dried, so no photographs were taken, and values of the coverage and complexity score were assumed to be zero. Sterile LB medium coverage and complexity score did not change from the unmodified values of about zero.

3.2.2 Modification with glucose

In samples that were modified with the addition of glucose, the ferning coverage remained consistent, while the complexity score changed from 5 to 8 at 2.5% and then held steady at higher concentrations of glucose (**Figure 5c**). The pattern on the plates transitioned from standard orthogonal ferning at low concentrations of glucose (**Figure 5c**: 0.5%, top left) to acute branching at high concentrations of glucose (**Figure 5c**: 4.5%, top right). The coverage and the complexity score on the sterile plate remained unchanged from the standard values.

3.2.3 Modification with sucrose

The desiccated plates of medium modified with sucrose had the most unusual patterns (**Figure 5d**). Both the coverage (47–30%) and the complexity score (5–3) dropped when the concentration of sucrose increased from 0 to 2%. However, with further increase from 2 to 4% sucrose, the coverage increased to 50%, while the complexity score continued to show less degrees of branching. Similar to glycerol, sucrose is hygroscopic, so plates appeared shinier and somewhat wet with increasing concentration of sucrose. At the same time, the ferning on the surface evolved (Table S1, https://ir.library.oregonstate.edu/concern/defaults/g158bp85b) from the orthogonal pattern (**Figure 5d**: 0.5%, top left) to fine strands to finally no ferning at high concentrations of sucrose (**Figure 5d**: 4.5%, top right). The glare in the photos from the shiny surface along with the packing of the fine crystalline strands produced high estimations of surface coverage until both the coverage and

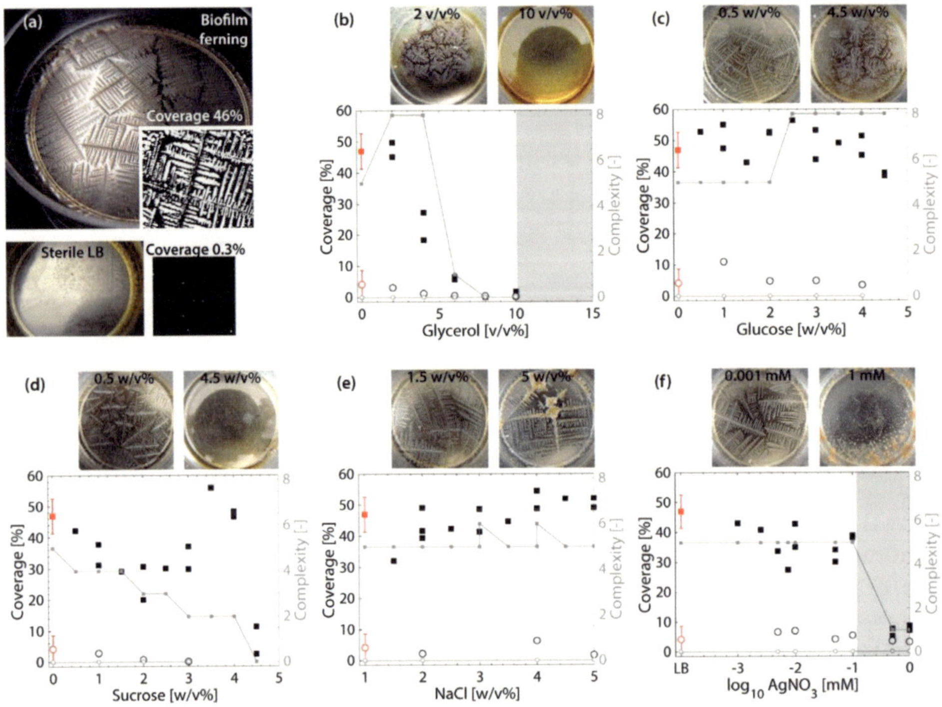

Figure 5.

(a) The original photograph was converted to a black and white image before the ferning coverage was calculated from the processed image. An example of ferns from biofilm that was grown in unmodified LB medium with 46.2% ferning coverage and from a plate of sterile unmodified LB medium showing zero ferning coverage. (b–f) Ferning coverage (left y-axis in black) and complexity score (right y-axis in gray) of biofilm that was grown in modified LB medium (coverage: black filled squares; complexity: Gray filled squares) and of sterile plates (coverage: black unfilled circles; complexity: Gray unfilled circles). The mean coverage and standard deviation of biofilm that was grown in unmodified LB medium (n = 12) and in sterile unmodified LB medium (n = 7) are plotted in red across figures (b–f). The results of the biofilm that was grown in modified LB medium with (b) glycerol, (c) glucose, (d) sucrose, (e) NaCl and (f) $AgNO_3$ are shown. The change in morphology of the biofilm ferning pattern with (b) glycerol at concentrations of 2% (top left) and 10% (top right); (c) glucose at concentrations of 0.5% (top left) and 4.5% (top right); (d) sucrose at concentrations of 0.5% (top left) and 4.5% (top right); (e) NaCl at concentrations of 1.5% (top left) and 5% (top right); and (f) $AgNO_3$ at concentrations of 0.001 mM (top left) and 1 mM (top right) are shown. The gray regions in (b) and (f) represent inhibiting concentrations that had no biofilm growth.

the complexity score suddenly dropped to zero at the maximum concentration of sucrose that was tested (4.5%). The complexity score of the pattern steadily dropped from 5 to 0 with sucrose, while the calculation of coverage resulted in scattered results, as the surface glare and the interaction of the densely packed linear striations with light affected the procedure.

3.2.4 Modification with sodium chloride

With further addition of sodium chloride (NaCl) to the modified LB medium, the ferning coverage remained consistently around 50%, and the complexity score remained around 5 (**Figure 5e**). Still, a change to the pattern exists, as the ferns evolved from thin branches (**Figure 5e**: 1.5%, top left) to a more pronounced branching with large crystalline formations with increasing concentration of NaCl (**Figure 5e**: 5%, top right). While the ferning branches became wider with the further addition of NaCl, the complexity score did not change. The orthogonal morphology and the non-overlapping crystallization appear to naturally limit the maximum coverage of the ferning pattern, resulting in a consistent 40–50% coverage. The sterile dishes with modified LB medium that contained the same amount of salt did not produce ferning patterns, so the coverage and complexity score remained zero.

3.2.5 Modification with silver nitrate

The complexity score for the plates that were treated with silver nitrate ($AgNO_3$) remained 5 for concentrations below 0.1 mM, but at higher concentrations, no biofilm growth occurred, resulting in a complexity score of 1 (**Figure 5f**). Even with no measurable biofilm, the plates contained clusters of dried materials that resembled nucleation points (**Figure 5f**: 1 mM, top right). For concentration below 0.1 mM of silver nitrate, the ferning patterns were orthogonal (**Figure 5f**: 0.001 mM, top left), and the coverage was in the range of 30–40%. Finally, the coverage dropped below 10% at concentrations that were greater than the inhibiting concentration of 0.1 mM. The coverage did not completely drop to 0, as beige clusters were left behind on the plate, which was the same reason that the complexity score was 1 for the highest concentration even though no biofilm was present.

3.2.6 Summary of ferning patterns

Similar to the dependence of rheological properties of the biofilm on the nutrient environment, the ferning pattern was dependent on the properties of the bacterial biofilm, so it changed with the composition of the medium. No ferning existed on plates that lacked biofilm. The presence of biofilm, confirmed rheologically and visually, correlated with robust ferning patterns. Using the same box-counting method, ferning patterns on the samples of unmodified LB biofilm had a fractal dimension of 1.8 (Figure S5, https://ir.library.oregonstate.edu/concern/defaults/g158bp85b), which is consistent with samples of mucus that had a fractal dimension of 1.7 [44]. The plates of sterile medium that were incubated and dried under the same conditions as the plates of biofilm had no visible ferning pattern, no yield stress, and little to no elastic modulus. So, the medium composition not only affected the growth of the biofilm and its rheological properties but also, by extension, affected the ferning pattern.

The complexity score and the ferning coverage was higher for stronger biofilms (higher G^* and τ_y) that caused more limitations in mass transfer, and both values dropped to nearly zero when no biofilm was present. The morphology of ferns with a complexity score of 8 that were produced by the biofilms with higher elasticity was similar to random/acute-angled branching ferns that were produced under conditions of increased diffusion limitation in a previous study [47]. Exceptions were the plates that appeared to never fully dry due to the high concentration of hygroscopic materials like sucrose or glycerol. So, even as the rheological properties of the biofilm increased (glycerol) or stayed constant (sucrose), the complexity score and coverage dropped with increasing amounts of the modifying chemical. The ferning coverage never exceeded 60%, indicating a natural growth limitation that was based on the available space and the morphology of the ferns. The videos of the ferning process demonstrated how these branches quickly started and stopped growing without any of the branches overlapping (Video S1, https://ir.library.oregonstate.edu/concern/defaults/g158bp85b).

The ferning patterns of the biofilms were large (visible without microscopy on the order of centimeters) with most of the patterns consisting of orthogonal branches, and the ferns were reproducible in coverage and complexity score. The ferning patterns that were formed by the biofilm had the same morphology as ferning patterns of saliva, cervical mucus, *E. coli*, salt-gel, or salt-protein [44, 47, 49, 50]. From reports in the literature and the results in this work, the orthogonal or oblique branching seemed to be the most common type of ferning with the examples of branches with acute angles being rare [47]. The cause of the change in the morphology of the fern in biofilms from 90° angles to acute angles is not immediately clear. However, other studies have reported that the gelatin-to-salt ratio was the key factor controlling the ferning morphology of salt-gelatin mixtures [47]. Therefore, the samples with higher rheological properties (glycerol and glucose samples), which arguably has a higher amount of EPS, may have produced branching with acute angles due to the increased EPS-to-salt ratio. Thus, the acute-angle morphology dominated when the biofilms had larger rheological values, indicating higher EPS-to-salt ratio, while orthogonal-branching morphology dominated at intermediate ratios with no ferning at extremely high or low values of the EPS-to-salt ratio.

3.3 Birefringence of bacterial biofilms

Analyzing the pre-desiccation biofilm that was grown in 5% NaCl under polarized filters showed the entire sample sample lit with birefringent strands (**Figure 6**). The Pf bacteriophages produced by *P. aeruginosa* are known to self-assemble into liquid crystals that exhibit birefringence [64]. The birefringent strands (black arrows) are 50–100 μm in length, and they are evenly distributed throughout the sample (**Figure 6a** and **b**). A magnified view of the strands revealed that each strand was a bundle of smaller strands that were surrounded by cell clusters (yellow arrows) and biofilm (**Figure 6c**). The bacteriophages are about 6 μm in length, so the bundle was likely composed of hundreds of individual Pf phages that were assembled into one strand [64]. These birefringent strand clusters did not exist in samples lacking biofilm. A plate with 10% NaCl LB medium formed sheets of salt crystals, no visible biofilm, and no birefringent strands (Figure S6, https://ir.library.oregonstate.edu/concern/defaults/g158bp85b).

The 5% NaCl sample that was previously analyzed (**Figure 6**) was desiccated (**Figure 7a**) and examined with and without a polarized filter (**Figure 7b–k**). The red lines outlined the specific regions of the fern under the microscope. The region outlined in the circle (**Figure 7b–e**) was the square crystalline structure on the plate. Without a filter this region showed a cubic structure with a length of 4–5 mm per side with a nucleation point in the center and thin diagonal lines running through it with cavernous voids coming from the sides of the structure (**Figure 7b**). The center of the ferning structure showed cuboid lattice-like patterns of growth emerging from the seed point (**Figure 7e**). Similarly, the previous study of the ferning pattern in a gelatin-salt mix revealed interlocking salt blocks with a length of 10–30 μm on a side that formed the backbone of the ferning structure [47].

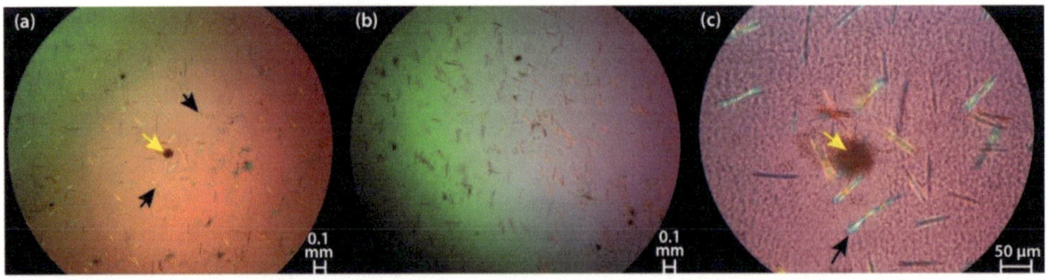

Figure 6.
Polarized microscopic images of inoculated samples of PAO1 in modified LB medium with a concentration of 5% NaCl. (a and b) Birefringent fragments exist throughout the liquid medium. (c) The birefringent thread-like fragments (black arrows) of about 50–100 μm in length were dispersed within medium that was full of bacterial clusters (yellow arrows).

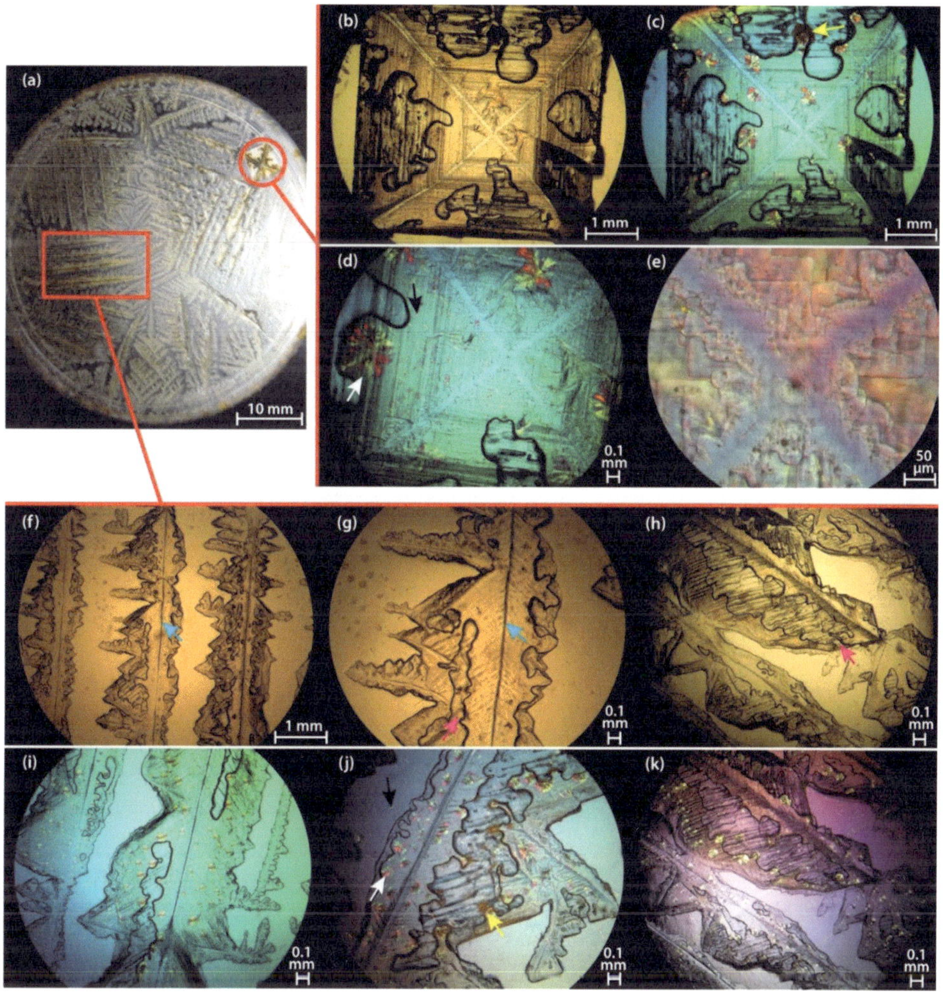

Figure 7.
Desiccated biofilm that was grown in LB medium with 5% NaCl. (a) Ferning pattern from the bottom of the petri dish. The regions that are outlined in red were inspected under the microscope. (b–e) The cubic piece at the top right corner of the plate (red circle): (b) seen under normal light; (c and d) seen through a polarized filter at different magnifications; (e) focused on the seed point of crystallization. (f and k) The ferning region of the plate (red square): (f–h) different regions of the fern under normal light at different magnifications; (i and j) seen through a polarized filter at different magnifications; (k) viewed at a different angle of polarization.

With a polarized filter, bright birefringent regions lit throughout the square (**Figure 7c**). A closer look at the center of the square revealed two forms of birefringent structures, large red and gold star formations (~0.5 mm in diameter) and red and gold strands that were about 50 μm in diameter (**Figure 7d**). In contrast to the birefringent strands that were scattered throughout the sample (black arrows), the birefringent stars (white arrows) were only in the crystallized ferns. Each birefringent strand that was visible in **Figure 7c** was a bundle of even smaller birefringent strands (**Figure 6c**). Therefore, the formation of the birefringent star demonstrated that having an even higher order of self-assembly during desiccation was possible such that the bundled strands further merged into a star formation. The alternating red and gold coloring indicated that the strands with matching orientations cluster together, but they must not have formed the entire cluster, as no star formation existed with only one color.

From the linear ferning section that was outlined by the red square (**Figure 7f–k**), the branches appeared to be about 1 mm in width with a distinct centerline running through each branch (**Figure 7f**). A magnified view of one of these branches revealed latticed or layered networks emanating from this central line (blue arrows) and cavities (pink arrows) that were present throughout the structure (**Figure 7g**). Some of the cavities were large, tunneling deep into the ferning structure (**Figure 7h**). Under polarized light, the branch was shown to have dozens of the star-shaped red and gold birefringent bundles (**Figure 7i**). Changing the angle of the polarized filter changed the color of the birefringent region from red and gold to gold and green (**Figure 7k**). The star-shaped birefringent clusters only existed within the crystal regions of the fern pattern, while the strands were scattered throughout the plate regardless of the ferning pattern (**Figure 7j**). This localization of the morphologies implied that the birefringent strands were produced within the biofilm; thus, they could be found throughout the material, while the formations of the birefringent stars were created as a result of crystallization, so they were only found within the crystalline regions. Clusters of bacterial cells appeared to be entrapped within the crystallized fern (yellow arrows), especially around the extremities of the ferning structure (**Figure 7c** and **j**). Similarly entrapped bacterial clusters were capable of reanimation at least a week after desiccation within the ferning structure [50]. Therefore, clusters of *P. aeruginosa* that were seen in **Figure 7c** and **j** may be in a suspended animation state as well, though this hypothesis was not tested during this study.

4. Conclusions

In environments that contained high viscosity (glycerol), high osmolarity (glycerol, NaCl), and high concentrations of simple carbon (glucose), the elasticity and the yield stress of the biofilm increased. Silver nitrate had an inhibiting effect on the biofilm formation, but only at concentrations that were greater than 0.1 mM. Similarly, concentrations of glycerol greater than 10% completely inhibited biofilm growth. However, the complex carbon structure of sucrose meant that it could not be utilized as an additional carbon source by PAO1 in the same way that glucose was utilized. Therefore, sucrose did not change the rheological properties of the biofilm. So, *P. aeruginosa* developed stronger biofilm under nutrient-rich conditions, certain levels of osmotic stress, and certain levels of diffusion limitation. However, it would not develop biofilm when the osmotic stress or diffusion limitation exceeded an inhibition amount or when an antimicrobial agent exceeded its inhibition concentration.

While the rheological properties of biofilm revealed information about the strength of the biofilm, the morphology of the ferning pattern best described the interactions between the electrolytes and the EPS in the biofilm. Typically, the biofilm had ferning coverage of about 50% and a ferning complexity score of 5. The ferning complexity increased with the strength of the biofilm (high complex modulus and yield stress), as stronger biofilm increased diffusion limitation that was experienced by the solutes within the matrix. The coverage and complexity score both dropped to zero when no biofilm formed, so the macromolecule-to-salt ratio was too low for ferning to occur, as with high concentrations of silver nitrate and glycerol. Many of the analysis methods of biofluid ferning patterns were qualitative and subjective, which is currently problematic considering its use as an indicator of certain medical symptoms. The image analysis and ferning classification method that was presented here could easily be applied to the other fields to give more quantitative values to the analysis of ferning biofluids.

The birefringence that was produced by liquid crystals within the samples of biofilm had two different morphologies, bundled strands that were about 50 μm in length in hydrated biofilm and star-shaped bundle of strands that were almost ten times larger inside the crystalline region of the ferning pattern. So, in addition to the self-assembly of the phages to strands inside the biofilm, a more complex assembly took place during crystallization in the biofilm that produced this tertiary structure. During the ferning process, clusters of bacteria became entrapped within the crystalline phase. Other researchers have found that these entrapped bacteria are in suspended animation state and that they could be brought back to life upon rehydration. If PAO1 can also reanimate, then ferning is yet another mechanism that *P. aeruginosa* could utilize to survive extreme conditions, similar to how liquid crystals formed by phages enhanced the resistance and persistence of *P. aeruginosa*.

Acknowledgements

The authors would like to thank Dr. Skip Rochefort for consulting on the rheology tests and Kristin Marshall for her help with the image conversion work. Additionally, the authors would like to thank Marisa Thierheimer, Curran Gahan, and Dalton Myas for helping with experimental preparation.

Conflict of interest

We do not have any conflict of interests to declare.

Funding

This work would not be possible without funding from Medical Research Foundation of Oregon.

Supplemental materials

The supplemental documents for this section may be found at: https://ir.library.oregonstate.edu/concern/defaults/g158bp85b.

Author details

Uranbileg Daalkhaijav[1], Angela L. Dunham[1] and Travis W. Walker[2*]

1 School of Chemical, Biological, and Environmental Engineering, Oregon State University, Corvallis, Oregon, USA

2 Department of Chemical and Biological Engineering, South Dakota School of Mines and Technology, Rapid City, South Dakota, USA

*Address all correspondence to: travis.walker@sdsmt.edu

IntechOpen

© 2019 The Author(s). Licensee IntechOpen. This chapter is distributed under the terms of the Creative Commons Attribution License (http://creativecommons.org/licenses/by/3.0), which permits unrestricted use, distribution, and reproduction in any medium, provided the original work is properly cited.

References

[1] Vu B, Chen M, Crawford RJ, Ivanova EP. Bacterial extracellular polysaccharides involved in biofilm formation. Molecules. 2009;**14**(7):2535-2554

[2] Lieleg O, Caldara M, Baumgärtel R, Ribbeck K. Mechanical robustness of *Pseudomonas aeruginosa* biofilms. Soft Matter. 2011;**7**(7):3307-3314

[3] Hall-Stoodley L, Costerton JW, Stoodley P. Bacterial biofilms: From the natural environment to infectious diseases. Nature Reviews Microbiology. 2004;**2**(2):95-108

[4] Flemming H-C, Wingender J. The biofilm matrix. Nature Reviews Microbiology. 2010;**8**(9):623-633

[5] Wloka M, Rehage H, Flemming H-C, Wingender J. Rheological properties of viscoelastic biofilm extracellular polymeric substances and comparison to the behavior of calcium alginate gels. Colloid & Polymer Science. 2004;**282**(10):1067-1076

[6] Yuan SJ, Pehkonen SO. Microbiologically influenced corrosion of 304 stainless steel by aerobic *Pseudomonas* NCIMB 2021 bacteria: AFM and XPS study. Colloids and Surfaces B: Biointerfaces. 2007;**59**:87-99

[7] Schuster M, Greenberg EP. A network of networks: Quorum-sensing gene regulation in *Pseudomonas aeruginosa*. International Journal of Medical Microbiology. 2006;**296**(2-3):73-81

[8] Silby MW, Winstanley C, Godfrey SAC, Levy SB, Jackson RW. Pseudomonas genomes: diverse and adaptable. FEMS Microbiology Reviews. 2011;**35**(4):652-680

[9] Barraud N, Hassett DJ, Hwang S-H, S a R, Kjelleberg S, Webb JS. Involvement of nitric oxide in biofilm dispersal of *Pseudomonas aeruginosa*. Journal of Bacteriology. 2006;**188**(21).7344-7353

[10] Palmer KL, Aye LM, Whiteley M. Nutritional cues control *Pseudomonas aeruginosa* multicellular behavior in cystic fibrosis sputum. Journal of Bacteriology. 2007;**189**(22):8079-8087

[11] Mowat E, Paterson S, Fothergill JL, Wright EA, Ledson MJ, Walshaw MJ, et al. *Pseudomonas aeruginosa* population diversity and turnover in cystic fibrosis chronic infections. American Journal of Respiratory and Critical Care Medicine. 2011;**183**(12):1674-1679

[12] Lau GW, Hassett DJ, Ran H, Kong F. The role of pyocyanin in *Pseudomonas aeruginosa* infection. Trends in Molecular Medicine. 2004;**10**(12):599-606

[13] Rada B, Leto TL. Pyocyanin effects on respiratory epithelium: Relevance in *Pseudomonas aeruginosa* airway infections. Trends in Microbiology. 2013;**21**(2):73-81

[14] Ghafoor A, Hay ID, Rehm BHA. Role of exopolysaccharides in *Pseudomonas aeruginosa* biofilm formation and architecture. Applied and Environmental Microbiology. 2011;**77**(15):5238-5246

[15] Folkesson A, Jelsbak L, Yang L, Johansen HK, Ciofu O, Høiby N, et al. Adaptation of *Pseudomonas aeruginosa* to the cystic fibrosis airway: An evolutionary perspective. Nature Reviews Microbiology. 2012;**10**(12):841-851

[16] Livermore DM. Multiple mechanisms of antimicrobial resistance in *Pseudomonas aeruginosa*: Our worst nightmare? Clinical Infectious Diseases. 2002;**34**(5):634-640

[17] Davies D. Understanding biofilm resistance to antibacterial agents.

Nature Reviews Drug Discovery. 2003;**2**(2):114-122

[18] Berk V, Fong JCN, Dempsey GT, Develioglu ON, Zhuang X, Liphardt J, et al. Molecular architecture and assembly principles of vibrio cholerae biofilms. Science. 2012;**337**(6091):236-239

[19] Bertani G. Lysogeny at mid-twentieth century: P1, P2, and other experimental systems. Journal of Bacteriology. 2004;**186**(3):595-600

[20] Acumedia. Yeast Extract (7184); 2011

[21] Bertani G. Studies on lysogenesis. I. The mode of phage liberation by lysogenic Escherichia coli. Journal of Bacteriology. 1951;**62**:293-300

[22] BD Biosciences. BD Bionutrients™ Technical Manual—Advanced Bioprocessing; 2006

[23] Organo Technie. Standard Series: Yeast Extract 19512; 2008

[24] De Kievit TR. Quorum sensing in *Pseudomonas aeruginosa* biofilms. Environmental Microbiology. 2009;**11**(2):279-288

[25] Shrout JD, Chopp DL, Just CL, Hentzer M, Givskov M, Parsek MR. The impact of quorum sensing and swarming motility on *Pseudomonas aeruginosa* biofilm formation is nutritionally conditional. Molecular Microbiology. 2006;**62**(5):1264-1277

[26] Schuster M, Sexton DJ, Diggle SP, Greenberg EP. Acyl-homoserine lactone quorum sensing: From evolution to application. Annual Review of Microbiology. 2013;**67**:43-63

[27] Shemesh M, Chaia Y. A combination of glycerol and manganese promotes biofilm formation in Bacillus subtilis via histidine kinase KinD signaling. Journal of Bacteriology. 2013;**195**(12):2747-2754

[28] Klausen M, Aaes-Jørgensen A, Molin S, Tolker-Nielsen T. Involvement of bacterial migration in the development of complex multicellular structures in *Pseudomonas aeruginosa* biofilms. Molecular Microbiology. 2003;**50**(1):61-68

[29] Lim Y, Jana M, Luong TT, Lee CY. Control of glucose-and NaCl-induced biofilm formation by rbf in Staphylococcus aureus. Journal of Bacteriology. 2004;**186**(3):722-729

[30] O'Toole GA, Kolter R. Initiation of biofilm formation in Pseudomonas fluorescens WCS365 proceeds via multiple, convergent signalling pathways: A genetic analysis. Molecular Microbiology. 1998;**28**(3):449-461

[31] Damkiær S, Yang L, Molin S, Jelsbak L. Evolutionary remodeling of global regulatory networks during long-term bacterial adaptation to human hosts. Proceedings of the National Academy of Sciences of the United States of America. 2013;**110**(19):7766-7771

[32] Ferreira C, Pereira AM, Melo LF, Simões M. Advances in industrial biofilm control with micro-nanotechnology. Current Research, Technology and Education Topics in Applied Microbiology and Microbial Biotechnology. 2010;**2**:845-854

[33] Raulio M. Ultrastructure of Biofilms Formed by Bacteria from Industrial Processes. Helsinki: University of Helsinki; 2010

[34] Bjarnsholt T, Kirketerp-Møller K, Kristiansen S, Phipps R, Nielsen AK, Jensen PØ, et al. Silver against *Pseudomonas aeruginosa* biofilms. APMIS. 2007;**115**(8):921-928

[35] Larson RG. The Structure and Rheology of Complex Fluids. New York, Oxford: Oxford University Press; 1999. 688 p

[36] Hamley IW. Introduction to Soft Matter (Revised Ed.). Introduction to Soft Matter: Synthetic and Biological Self-Assembling Materials. Chichester, UK: John Wiley & Sons, Ltd; 2007. pp. 1-328

[37] Pavlovsky L, Younger JG, Solomon MJ. In situ rheology of Staphylococcus epidermidis bacterial biofilms. Soft Matter. 2013;**9**(1):122

[38] Towler BW, Rupp CJ, Cunningham AB, Stoodley P. Viscoelastic properties of a mixed culture biofilm from rheometer creep analysis. Biofouling. 2003;**19**(5):279-285

[39] Vinogradov AM, Winston M, Rupp CJ, Stoodley P. Rheology of biofilms formed from the dental plaque pathogen Streptococcus mutans. Biofilms. 2004;**1**(1):49-56

[40] Kovach K, Davis-Fields M, Irie Y, Jain K, Doorwar S, Vuong K, et al. Evolutionary adaptations of biofilms infecting cystic fibrosis lungs promote mechanical toughness by adjusting polysaccharide production. npj Biofilms and Microbiomes. 2017;**3**(1):1

[41] Shaw T, Winston M, Rupp CJ, Klapper I, Stoodley P. Commonality of elastic relaxation times in biofilms. Physical Review Letters. 2004;**93**(9):098102

[42] Houari A, Picard J, Habarou H, Galas L, Vaudry H, Heim V, et al. Rheology of biofilms formed at the surface of NF membranes in a drinking water production unit. Biofouling. 2008;**24**(4):235-240

[43] Daalkhaijav U, Walker TW. Developing a nondestructive technique for measuring bulk rheology of p*seudomonas Aeruginosa* biofilm. Applied Rheology. 2017;**27**(6):64033-64043

[44] Cortés ME, Hauyón R, González F, Vigil P. Evidence of fractality in a pattern of crystallization of bovine cervical mucus obtained at oestrus. International Journal of Morphology. 2012;**30**(4):1461-1465

[45] Cortés ME, González F, Vigil P. Crystallization of bovine cervical mucus at oestrus: An update. Revista de Medicina Veterinaria. 2014;**28**:103-116

[46] Losa GA. The fractal geometry of life. Rivista di Biologia-Biology Forum. 2009;**102**(1):29-60

[47] Goto M, Oaki Y, Imai H. Dendritic growth of NaCl crystals in a gel matrix: Variation of branching and control of bending. Crystal Growth & Design. 2016;**16**(8):4278-4284

[48] Maragou M, Vaikousis E, Ntre A, Koronis N, Georgiou P, Hatzidimitriou E, et al. Tear and saliva ferning tests in Sjögren's syndrome (SS). Clinical Rheumatology. 1996;**15**(2):125-132

[49] Kogbe O, Liotet S, Tiffany JJM. Factors responsible for tear Ferning. Cornea. 1991;**10**(5):433-434

[50] Gómez Gómez JM, Medina J, Hochberg D, Mateo-Martí E, Martinez-Frias J, Rull F. Drying bacterial biosaline patterns capable of vital reanimation upon rehydration: Novel hibernating biomineralogical life formations. Astrobiology. 2014;**14**(7):589-602

[51] Desarnaud J, Derluyn H, Carmeliet J, Bonn D, Shahidzadeh N. Metastability limit for the nucleation of NaCl crystals in confinement. Journal of Physical Chemistry Letters. 2014;**5**(5):890-895

[52] Tabbara KF, Okumoto M. Ocular ferning test. A qualitative test for mucus deficiency. Ophthalmology. 1982;**89**(6):712-714

[53] Rolando M, Baldi F, Zingirian M. The effect of hyperosmolarity on tear mucus ferning. Fortschritte der Ophthalmologie. 1986;**83**(6):644-646

[54] Ravinder R, Kaipa O, Baddela VS, Singhal Sinha E, Singh P, Nayan V, et al. Saliva ferning, an unorthodox estrus detection method in water buffaloes (*Bubalus bubalis*). Theriogenology. 2016;**86**(5):1147-1155

[55] Menárguez M, Pastor LM, Odeblad E. Morphological characterization of different human cervical mucus types using light and scanning electron microscopy. Human Reproduction. 2003;**18**(9):1782-1789

[56] Gómez Gómez JM, Medina J, Rull F. A rich morphological diversity of biosaline drying patterns is generated by different bacterial species, different salts and concentrations: Astrobiological implications. Astrobiology. 2016;**16**(7):513-524

[57] Arakaki A, Shimizu K, Oda M, Sakamoto T, Nishimura T, Kato T. Biomineralization-inspired synthesis of functional organic/inorganic hybrid materials: Organic molecular control of self-organization of hybrids. Organic & Biomolecular Chemistry. 2015;**13**(4):974-989

[58] Li X, Chopp DL, Russin WA, Brannon PT, Parsek MR, Packman AI. Spatial patterns of carbonate biomineralization in biofilms. Applied and Environmental Microbiology. 2015;**81**(21):7403-7410

[59] Thorkelsson K, Bai P, Xu T. Self-assembly and applications of anisotropic nanomaterials: A review. Nano Today. 2015;**10**(1):48-66

[60] Hamley IW. Liquid crystal phase formation by biopolymers. Soft Matter. 2010;**6**(9):1863-1871

[61] Araki J, Kuga S. Effect of trace electrolyte on liquid crystal type of cellulose microcrystals. Langmuir. 2001;**17**(15):4493-4496

[62] Secor PR, Jennings LK, Michaels LA, Sweere JM, Singh PK, Parks WC, et al. Biofilm assembly becomes crystal clear—Filamentous bacteriophage organize the *Pseudomonas aeruginosa* biofilm matrix into a liquid crystal. Microbial Cell. 2016;**3**(1):49-52

[63] Unwin RR, Cabanas RA, Yanagishima T, Blower TR, Takahashi H, Salmond GPC, et al. DNA driven self-assembly of micron-sized rods using DNA-grafted bacteriophage fd virions. Physical Chemistry Chemical Physics. 2015;**17**(12):8194-8202

[64] Secor PR, Sweere JM, Michaels LA, Malkovskiy AV, Lazzareschi D, Katznelson E, et al. Filamentous bacteriophage promote biofilm assembly and function. Cell Host & Microbe. 2015;**18**(5):549-559

[65] Tomar S, Green MM, Day LA. DNA-protein interactions as the source of large-length-scale chirality evident in the liquid crystal behavior of filamentous bacteriophages. Journal of the American Chemical Society. 2007;**129**(11):3367-3375

[66] Secor PR, Michaels LA, Smigiel KS, Rohani MG, Jennings LK, Hisert KB, et al. Filamentous bacteriophage produced by *Pseudomonas aeruginosa* alters the inflammatory response and promotes noninvasive infection in vivo. Infection and Immunity. 2017;**85**(1):1-11

[67] Penner JC, Ferreira JAG, Secor PR, Sweere JM, Birukova MK, Joubert L, et al. Pf4 bacteriophage produced by *Pseudomonas aeruginosa* inhibits Aspergillus fumigatus metabolism via iron sequestration. Microbiology. 2016;**162**(9):1583-1594

[68] Vallooran JJ, Handschin S, Pillai SM, Vetter BN, Rusch S, Beck HP, et al. Lipidic cubic phases as a versatile

platform for the rapid detection of biomarkers, viruses, bacteria, and parasites. Advanced Functional Materials. 2016;**26**(2):181-190

[69] Stewart PS, Franklin MJ. Physiological heterogeneity in biofilms. Nature Reviews Microbiology. 2008;**6**(3):199-210

[70] Freitas F, Alves VD, Pais J, Costa N, Oliveira C, Mafra L, et al. Characterization of an extracellular polysaccharide produced by a pseudomonas strain grown on glycerol. Bioresource Technology. 2009;**100**(2):859-865

[71] Kim JS, Kuk E, Yu KN, Kim J-H, Park SJ, Lee HJ, et al. Antimicrobial effects of silver nanoparticles. Nanomedicine. 2007;**3**(1):95-101

[72] Mezger TG. The Rheology Handbook. 4th ed. Vincentz Network GmbH & Co KG. Hanover; 2014. 434 p

Chapter 3

Host-Pathogen Interaction in the Lung of Patients Infected with *Pseudomonas aeruginosa*

Sandra Grumelli

Abstract

Pseudomonas aeruginosa is an opportunistic bacterium that can proliferate in the soil, water, and even humans if they are immunologically depressed. During lung infections, *P. aeruginosa* goes through significant morphological changes turning into the mucoid form after which its eradication becomes almost impossible. Within this chapter, we explore the bioenergetics changes produced within *P. aeruginosa* during infections in humans and the metabolic pathways that are involved in those changes that lead to chronic infection.

Keywords: *P. aeruginosa*, host, bioenergetics, phosphate, choline

1. Introduction

There are many lung pathogens but one of the most studied is *Pseudomonas aeruginosa* because it cannot be eradicated under certain conditions. As an opportunistic pathogen, its interaction with the host has some particularities that we will explore in this chapter.

The *Pseudomonadaceae* comprise Gram-negative microorganism, nonsporulated, aerobic strict of wide distribution in the environment from the soil, water, and plants to humans; this is due to their nutritional versatility. Of this vast group, only *Pseudomonas cepacia*, *mallei*, and *aeruginosa* infect humans, of which *aeruginosa* is the more relevant because it is the most frequent cause of nosocomial infections [1].

It is often said that *Pseudomonas aeruginosa* does not infect healthy individuals but there are reports on the contrary, as swimmers otitis [2]. Because it is an opportunist pathogen, it does not need the host for its survival, and it may be lethal after becoming a chronic infection in susceptible patients with cystic fibrosis (CF) [3–5], cancer [6–8], hepatic cirrhosis [9], keratitis [10–13], or spondylodiscitis [14]. This bacterium is most feared by pulmonologist because when acquired by nosocomial patients [15, 16], it complicates any existing conditions, and when it invades immune-compromised patients, its eradication may become impossible.

Colonization with *P. aeruginosa* is observed in all stages of chronic obstructive pulmonary disease (COPD), but the prevalence significantly increases with disease severity from 0.7%, in stage 1 of the Global Initiative for Obstructive Lung Disease, to 1.5% for stages 2 and 3 up to 2.6% for stage 4 [17, 18]. This prevalence rises to 8–13% in acute exacerbations of COPD [19–21]. But still, the main susceptibility for the infection and death by *P. aeruginosa* [22, 23] are the mutations of the CF

IntechOpen

transmembrane conductance regulator (CFTR) identified as F508, G542X, G551D, W1282X, R1162X, and N1303K [24, 25]. CF also has co-morbidity such as liver cirrhosis [26] with 18% prevalence [27, 28] of *P. aeruginosa* infection in this subset.

2. Host-bacteria interaction in acute infection

2.1 Lung changes upon bacterial invasion

The flagella and lipopolysaccharide (LPS) from *P. aeruginosa* are the first to contact the ciliated epithelial cells [29]. In the airways, these cells are covered by the surfactants containing 45% less NaCl and 600 more K^+ than in plasma [30], while the alveolar epithelial cells are covered by a surfactant layer that contains mostly phosphatidylcholine (80%) [31] and surfactant proteins A, B, C, and D [32, 33] that bind LPS in a calcium-dependent manner [34]. After the surfactant layer is crossed, the flagellum binds to the epithelial cells through toll-like receptors (TLR) 2, 3, 4 and 5 [35–40] that are quickly endocytosed to be degraded in the proteasome. The activated TLR5 induces the macrophages chemoattractants CXCL1, CXCL2, and neutrophil chemokine CCL20, which are inhibited by TLR5 inhibitors [41]. The peptides digested are then presented to macrophages and dendritic cells.

When LPS binds to the host cells, where CFTR is also a receptor [42], it upregulates NF-κB at the gene level (**Table 1**), promoting inflammation [43] by secretion of IL1, IL6, IL8, ICAM-1, and also CXCL1 [44–47], although in different degrees of regulation. For example, CXCL1 expression is orchestrated by a fatty acid-binding protein (FABP4) that delivers fatty acids from the cytoplasm to the nuclear receptor PPAR. These prompt macrophage signaling through the myeloid differentiation protein-88 (MyD88) to induce cytokine production following engagement of TLRs with LPS [48–51]. Macrophages require MyD88 to produce CXCL1 but also eicosapentaenoic acid and docosahexaenoic acid, both substrates of FABP4. This demonstrates the importance of fatty acid metabolism to promote host resistance to *P. aeruginosa*, facilitating macrophage-neutrophil cross-talk during the infection [52, 53].

The T cells also play an important role in acute infection. IL17 producing T cells are expanded [54], via expression of STAT3 and retinoid orphan receptor [55]; these steps are crucial for B cell activation and immunoglobulin release for bacterial clearance [56]. On the contrary, excess of T regulatory cells (Treg) are associated with secondary *P. aeruginosa* infections, because depletion of Tregs decreases IL-10 levels and elevates IL-17A, IL-1β, and IL-6 [57, 58]. Therefore, the underlying immune suppression, by Treg accumulation, and Th17 depletion are the cause of chronic infection [57]. This may be reversed by treatment with IL7 or ethyl pyruvate increasing IL17, INFγ, and $CD8^+$ T cells [59, 60].

Death of CF patients chronically infected with *P. aeruginosa* occurs due to the depletion of neutrophils, IL6, and granulocyte-colony stimulating factor which causes dysfunctional neutrophil burst. This reduces the secretion of reactive oxygen species, which are essential for bacterial killing and clearance [61].

2.2 Bacterial metabolic changes for invasion

Simultaneously, the contact of *P. aeruginosa* with the lung upregulates in the bacteria genes involved mainly in biofilm synthesis [62] (**Table 1**). These changes in gene expression result in downregulation of proteins involved in LPS biosynthesis, antimicrobial resistance, and phenazine production concomitant

P. aeruginosa			**Lung**		
Gene ID	**FC**	**Name**	**Gene ID**	**FC**	**Name**
hemE	16.1[a]	Uroporphyrinogen decarboxylase		4502	Dioxin-inducible cytochrome P
pyrC	12.1	Dihydroorotase (biofilm development)		252	ppGpp
pyrH	6.4	Uridylate kinase (biofilm development)		206.7	Tumor necrosis factor-α-inducible DNA-binding protein A
adhA	5.5	Alcohol dehydrogenase		133.1	Proteasome subunit C
pelB, pelE	5.6, 3.7	Extracellular polysaccharide (competitive disruption of *S. aureus* biofilms)	hORC2L	13.4[b]	Human origin recognition complex protein 2
cls	7.0	Cardiolipin synthase	MCP-1	13.3	Monocyte chemotactic protein 1
pscD	3.2	T3SS export protein		3.5	c-Jun
plcN	3.2	Phospholipase C precursor		3	GTP-binding protein rhoB4.
algD,E,F,8,amrZ	1.9–10.7	Alginate biosynthesis		2.9	Urokinase-type plasminogen activator
ppiA	2.5	Peptidyl-prolyl cis-trans isomerase	PKC	2.8	Protein kinase C, ETA type
hmgA	−7.2	Homogentisate 1–2-dioxygenase		2.7	Folylpolyglutamate synthetase
algC	−9.3	Phosphomannomutase	TTP	5.7	Tristetraproline
hemE	16.1	Uroporphyrinogen decarboxylase		2.4	Anti-oncogene
pyrC	12.1	Dihydroorotase (biofilm development)	MAD3	5.1	IκB-α
pyrH	6.4	Uridylate kinase (biofilm development)	hENT1	4.2	Placental equilibrative nucleoside transporter 1
adhA	5.5	Alcohol dehydrogenase	TEL	2.8	Transcription factor
pelB, pelE	5.6, 3.7	Extracellular polysaccharide (competitive disruption of *S. aureus* biofilms)	DPH2L	2.6	Diphtheria toxin resistance protein
cls	7.0	Cardiolipin synthase	TFPI2	2.3	Tissue factor pathway inhibitor 2
				2.1	Ankyrin motif
			ESE-1	2.1	Epithelial-specific transcription factor
				−12.5	8IRF
				−11.9	JAK-1
			EPB49	−2.0	Erythrocyte membrane protein band 4.9. (Dematin)
				−2.3	*Alu* repeat-containing sequence

FC, fold changes; NC, no change.
[a]Change relative to P. aeruginosa acute infection/chronic contact to host cell [62].
[b]Relative change of lung cell gene profile after 3 h contact with P. aeruginosa [43].
Data reported by Naughton [62] and Ichikawa et al. [43].

Table 1.
Genetic changes due to host-pathogen interaction quantified by microarrays of mRNA.

with the upregulation of proteins involved in adherence, lysozyme resistance, and inhibition of the chloride ion channel, and CFTR [63]. *P. aeruginosa* releases choline from surfactants [81]. *In vitro* studies utilizing choline, as a carbon and nitrogen source, shows that it produces accumulation of polyphosphates (polyPi), carbohydrates, and LPS accompanied by depletion of phosphate (Pi) and phospholipids (PL); deeply modifying its energetic metabolism, the bacteria save 45% of energy in polyPi [64] (**Table 2**).

After the invasion, the bacteria attach to the lung epithelium producing profound metabolic changes, which correlates with morphological changes to the rugose small-colony variant (RSCV) [65–67]. The transition to the RSCV precedes inactivation of serine hydroxymethyltransferase; this produces accumulation of cyclic diguanylate [68] and nucleotide ppGpp that leads to polyPi accumulation [69] and to alginate production [68, 70–72].

Table 2 shows that the total content of phosphate is reduced 3 times in choline feed bacteria, although it accumulates Pi in polyPi. The polyPi may be thought as the energetic savings of the bacteria which is done at expenses of phospholipid biosynthesis. This is possible reducing the size of the bacterium [73] and increasing the area/volume ratio that facilitates O_2 exchange for which the bacteria have to compete with the host [74]. The overall bacterial changes save energy accumulating ppGpp, the substrate for polyPi synthesis by polyphosphate kinase, which is also increased [75]. Some of these polyPi are located in the outer membrane where this highly energetic polymer has Pi bonds similar to the ATP and a highly negative charge neutralized by cations such as Ca^{2+} and Mg^{2+}. Thus, polyPi function as an energy storage, buffer, and ion chelator that may shield the bacterium from environmental changes.

After adhering to the host ciliated epithelial cells, through mucin, the bacterium is enabled to form aggregates, secrete alginate, and modify its LPS [76]; this is a process regulated by 3,5-cyclic diguanylic acid [68]. The LPS is a macromolecule

Composition	Succinate[a] + NH_4Cl		Choline[a]			
	μg/mg of protein	μmol/mg of protein	μg/mg of protein	μmol/mg of protein	%	p[b]
Phosphate	1400 ± 100	14.7 ± 0.7	460 ± 90	4.8 ± 0.7	33	0.001
ATP	1650 ± 330	3.0 ± 0.6	1270 ± 165	2.3 ± 0.3	−23	0.32
Polyphosphates	4.0 ± 1.8	0.042 ± 0.01	6.3 ± 1.4	0.066 ± 0.008	57	0.004
Carbohydrates[c]	210 ± 40	1.2 ± 0.2	330 ± 50	1.8 ± 0.2	50	0.03
LPS[d]	19 ± 4	0.08 ± 0.02	41 ± 9	0.16 ± 0.03	100	0.02
Phospholipids[e]	114 ± 7	0.65 ± 0.04	71 ± 4	0.1 ± 0.02	−85	
Biosynthetic energy (ATP)[f]	—	1675	—	924	45	

[a]*Bacteria were grown in a high phosphate basal salt medium. All chemical determinations were done on 1.05 ± 0.16 and 1.00 ± 0.20 mg ml* $^{-1}$ *of culture from whole bacteria grown with 20 mM succinate plus 18.7 mM* NH_4C1 *or 20 mM choline chloride, respectively. Results are the average of four independent experiments ± SD.*
[b]*Values obtained by ANOVA analysis.*
[c]*Total carbohydrates were measured by the phenol method.*
[d]*Measured as the content of KDO according to the determination of formylpyruvic with thiobarbituric acid.*
[e]*Total phospholipids from bacteria grown with succinate/NH4Cl or choline.*
[f]*Value obtained by calculation of the biosynthetic cost of LPS 470 μmol ATP/gr of cells, 1 μmol ATP/g polyphosphate, 470 μmol ATP/g of glycoside, and 2578 μmol ATP/g of phospholipids. Table taken from Grumelli [64].*

Table 2.
Metabolic changes in the bacteria upon infection.

DOI: http://dx.doi.org/10.5772/intechopen.84657

($C_{205}H_{366}N_3O_{117}P_5$) of 4899.956 g/mol that covers the outer membrane extending 40 nm outward. It is released with vesicle-containing enzymes and outer membrane (OM). Its extended formula was determined in 2003 (**Figure 1**); it is anchored to the OM through the lipid A which binds to the 3-deoxy-D-manno-2-octulosonic acid (KDO), the first glycoside of the core oligosaccharide, bound to the distal O antigen, a highly variable region [77, 78]. A metabolic crossroad between the LPS and alginate biosynthesis (**Figure 2**) is mannose-6-phosphate isomerization to mannose-1-phosphate by phosphomannomutase (Alg C). The glucose-6 phosphate (G6P) can be transformed to G1P to produce LPS or to isomerize mannose-6-phosphate to G1P. Similarly, fructose-6-phosphate (F6P) can be converted to mannose-6-phosphate and then isomerized to mannose-1-phosphate that becomes alginate by D-mannuronate linkage to L-guluronate via a P-1,4 glycosidic bond. Thus, isomerization of mannose 6-phosphate to mannose 1-phosphate by phosphomannomutase, encoded as algC, is common to the biosynthesis of LPS and alginate since mutants in this phosphomannomutase are hindered in their ability to infect *in vivo* [79].

2.3 Interaction between lung and bacteria

The host-pathogen interaction studied *in vivo* utilizing LPS in the lung of mice exposed to cigarette smoke model exacerbations of COPD in patients chronically infected with *P. aeruginosa*. **Figure 3** proposes that this extracellular pathogen releases to the medium phospholipase C (PLC) [80] and phosphorylcholine phosphatase (PChP) [81] within vesicles [82]. These vesicles degrade the surfactant, from phosphatidylcholine [85] to phosphoryl-choline and diacylglycerol (DAG) [83], causing Ca^{2+} mediated vaso-constriction [84]. Choline and phosphate (Pi) released by PChP produce airway constriction and inflammation in the lung tissue.

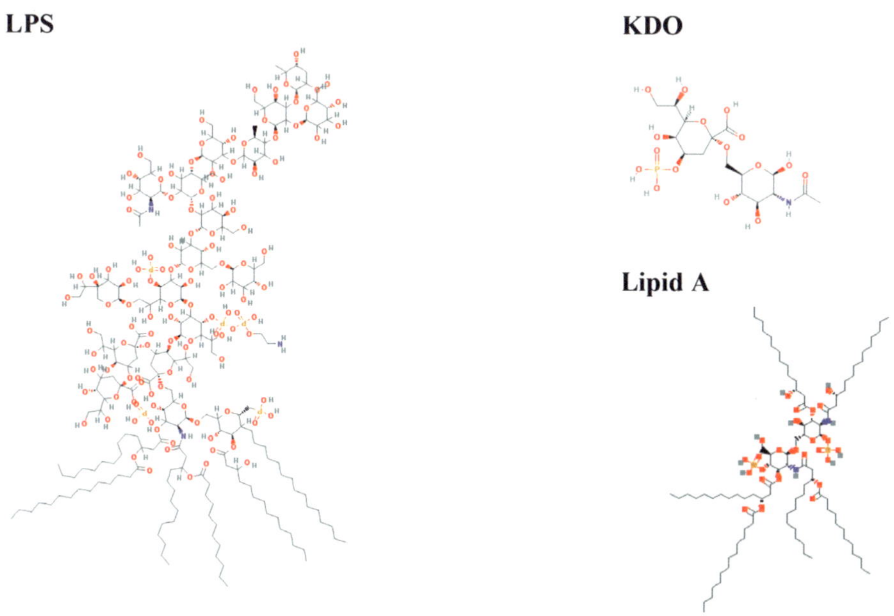

Figure 1.
LPS formula and structure set forth in PubChem (CID 11970143); and its parts KDO, (CID 49792052); and Lipid A (CID 9877306).

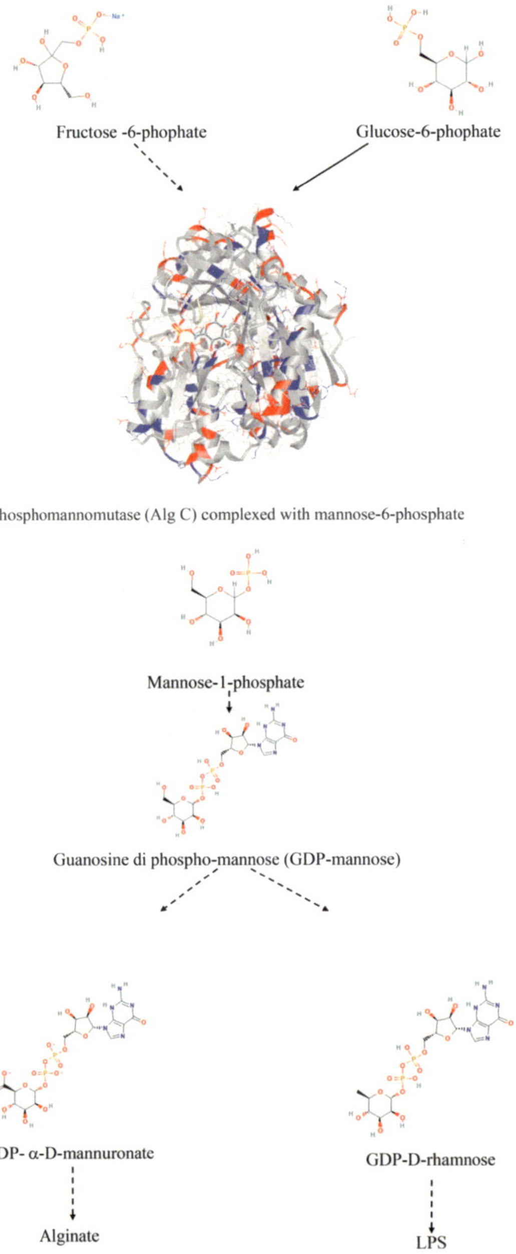

Figure 2.
The metabolic fork that derives glucose-6 phosphate (G6P) from biosynthesis of LPS to alginate. Tridimensional structure of phosphomannomutase; red and blue represent oppositely charged regions.

Further validation of this host-pathogen interaction is verified by the metabolite variations in a mouse model that uses live bacteria, instead of LPS. **Figure 4A** shows that phosphatidylcholine and glycine are significantly reduced in the lung upon infection, due to their consumption, while succinate and lactate are significantly accumulated [85]. Variations of choline concentration in the lung are not significant although glycerophosphocholine and glycine are [86, 87], which are the degradation

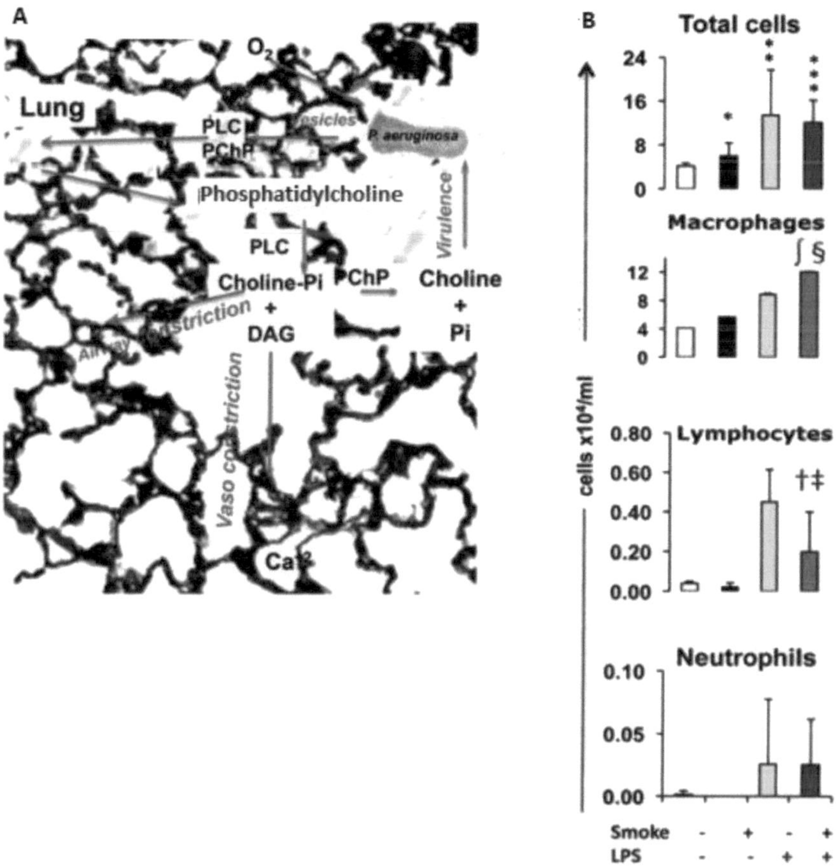

Figure 3.
*(A) Representative scheme of the host-pathogen interaction in mice lung during exacerbations of COPD. As an extracellular pathogen, P. aeruginosa releases to the medium phospholipase C (PLC) and phosphorylcholine phosphatase (PChP) within vesicles that degrades the membranes and surfactant of lung epithelial cells from phosphatidylcholine to phosphorylcholine and diacylglycerol (DAG) that cause Ca_2^+ mediated vaso-constriction. Choline and Pi released by PChP produces airway constriction in the lung tissue, and LPS and PolyPi accumulation in P. aeruginosa. (B) Representative experiment of inflammatory cells present in BAL of naïve mice (n = 5), mice treated with of LPS (n = 4), smoke exposed (n = 8) and smoke plus 100 ng/weekly of LPS (n = 3) from P. aeruginosa. *P = 0.01 relative to naïve mice, **P = 0.04 relative to naïve mice, ***P = 0.01 relative to smoke exposed, §P = 0.01 relative to naïve mice, †P = 0.05 relative to smoke exposed, ∫P = 0.05 relative to naïve mice, and ‡P = 0.01 relative to smoke exposed. The figure is taken from Grumelli et al. [64].*

products of choline. This is because *P. aeruginosa* is capable of releasing choline and converting it to betaine and then to glycine (**Figure 4B**) [88–91], for osmo-protection [92, 93] from the hyperosmolarity in the CF lung. Glycine also triggers chloride influx, inhibiting the Ca^{2+} mobilized by LPS [94]. This is a mechanism of self-preservation because macrophages are activated by LPS but suppressed by free glycine [95].

The succinate accumulated in the lung after infection [85], as Krebs cycle metabolite, inhibits histone demethylases, collagen hydrolases, α-ketoglutarate dioxygenases, and the 5-methylcytosine hydroxylase family [96]. *In vitro* succinate is the favorite carbon source for *P. aeruginosa*. Its consumption reduces the length of the LPS (**Table 3**), increasing the PL and Pi content and preventing the polyPi accumulation (**Table 2**), which is essential to the stress response [64]. The LPS and

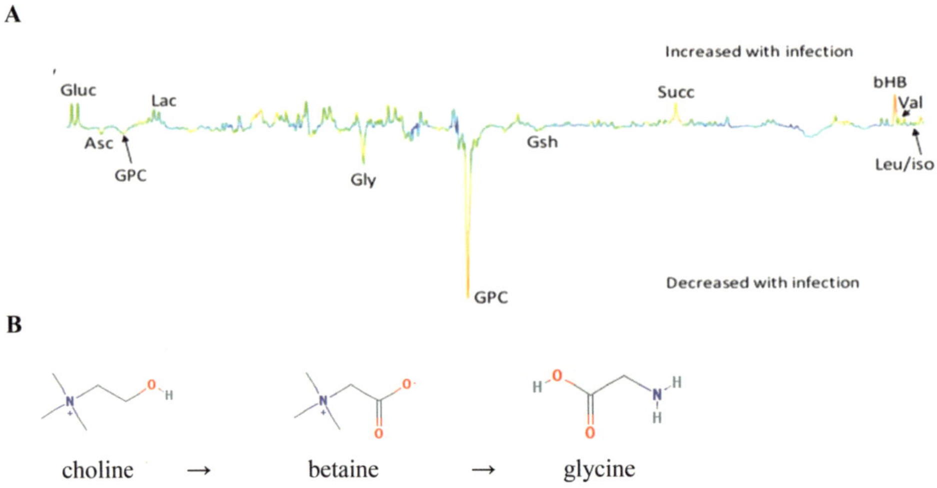

Figure 4.
(A) Lung alterations due to host-pathogen interaction upon infection. Gluc, glucose; Asc, ascorbate; GPC, glycerophosphocholine; Gly, glycine; Succ, succinate; bHB, beta-hydroxybutyrate; Val, valine; Leu/iso, leucine/isoleucine; Lac, lactate; and Gsh, glutathione reduced; figure taken from [85] and (B) choline conversion by P. aeruginosa.

Composition	Succinate[a] + NH_4 (µmol/µmol KDO)	Choline[a] (µmol/µmol KDO)	%	p
LPS[b]				
Total Pi	27 ± 5	33 ± 8	22	NS
Carbohydrates[c]	0.09 ± 0.01	0.15 ± 0.02	67	≤0.05
Lipid A				
Palmitic ac.[d]	34 ± 2	39 ± 5	15	NS[e]
12 carbon-hydroxyl ac.	32 ± 14	45 ± 20	41	NS

[a]Bacteria were grown in a high phosphate basal salt medium with 20 mM succinate plus 18.7 mM NH_4C1 or 20 mM choline chloride. All chemical determinations were carried out on LPS isolated with Triton X-100 from whole bacteria harvested at absorbance at 660 nm of 0.7. Total cellular contents were 1.05 + 0.16 and 1.00 + 0.20 mg/ml for succinate and choline, respectively. Results are the average of four independent experiments ± SD. P values were obtained by ANOVA analysis.
[b]KDO quantified.
[c]Carbohydrates quantified by the phenol method.
[d]Lipids were hydrolyzed from lipid A, identified by mass spectrometry. Results are expressed relative to stearic acid and averaged of three independent experiments ± SD.
[e]No significative. Data taken from Grumelli [64].

Table 3.
Variation in LPS composition according to the lung environmental changes.

PL biosynthesis has a common metabolite, the *R*-3-hydroxyacyl-ACP that is the substrate for *R*-3-hydroxyacyl-ACP dehydrase (FabZ) [98], to synthesize PL, and for LpxA, for LPS synthesis. Thus, the increased content of PL is at the expense of Lipid A from LPS (**Figure 5**), as shown in **Table 2**.

The LPS of *P. aeruginosa* stimulates the O_2 uptake from mitochondria [97] producing decoupling of the oxidative phosphorylation, reducing the respiratory rate, which generates stress in the host lung triggering exacerbations [44, 64, 97]. Therefore, succinate accumulation signifies that choline consumption is increasing the adaptation of the bacteria to the lung environment and the transition to the RSVC form, for chronic infection.

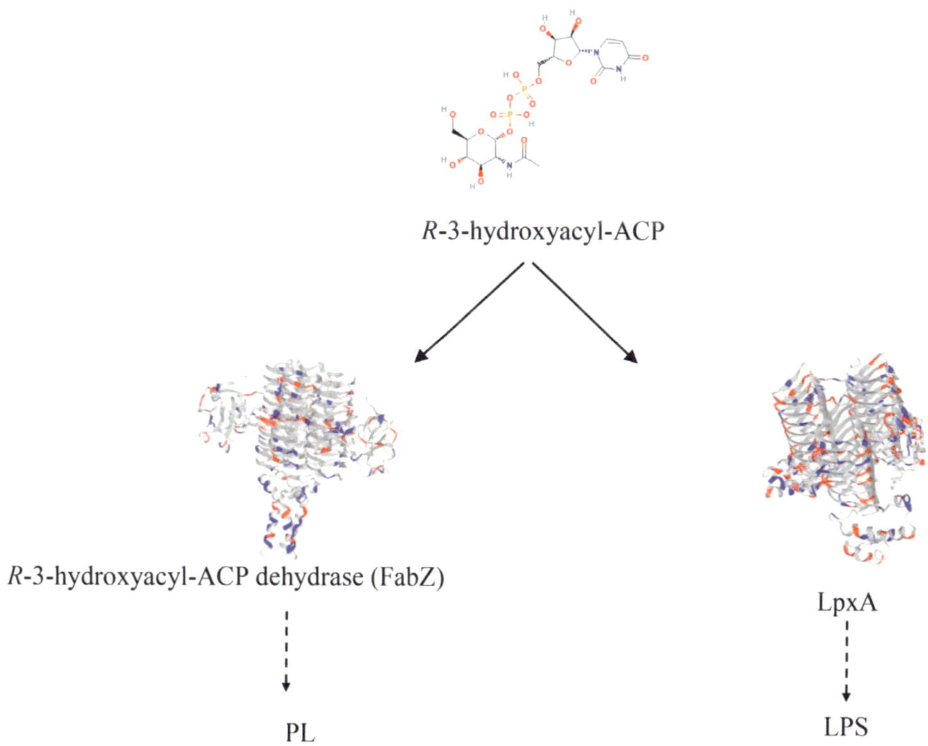

Figure 5.
R-3-hydroxyacyl-ACP, metabolite common to the biosynthesis of LPS and PL for which R-3-hydroxyacyl-ACP dehydrase (FabZ) and LpxA compete [98].

3. Chronic infection of *P. aeruginosa*

Upon infection, the host decreases iron levels in the blood [99]; this iron deficiency regulates a great number of bacterial virulent genes like alginate, the most relevant virulence factor, for *P. aeruginosa* survival [100]. In the lung, iron deficiency turns on AlgQ, the bacterial biofilm production gene, also known as AlgR2 [101, 102], under the Pfr A regulation that assists to the formation of two kinds of cytoplasmic aggregates: large vacuole-like bodies and smaller granules containing iron in association with oxygen or phosphate, very likely polyPi [103]. This leads to the RSCV type of *P. aeruginosa*. Under these conditions, the bacteria secrete alginate, a linear polysaccharide of D-mannuronic acid linked to L-guluronic acid [104].

The first gene described for the biosynthesis of alginate was the phosphomannose isomerase and GDP-mannose dehydrogenase (AlgD) that catalyze the conversion of GDP-mannose to GDP-mannuronic acid [105]. Upon oxygen limitation, *P. aeruginosa* utilizes nitrate or arginine as electron acceptors, via the succinylarginine pathway [106, 107]. The AlgD expression is tightly regulated by several environmental sources including nitrogen, O_2, Pi, NaCl, etc. Although the regulation of AlgD has been extensively studied, it is not completely understood, and eradication of chronic infection greatly depends on control of alginate production.

Several authors have studied the AlgD regulation, **Figure 6** shows a 20-years breach in the finding of AlgD regulators. More positive regulators have been identified, such as AlgR that is upregulated by NaCl and also by the nitrogen source [108]. AlgD is also under the same promoter than PLC, which is sensitive to the nitrogen source [109] that regulates the anaerobiosis genes. These genes detect the ratio

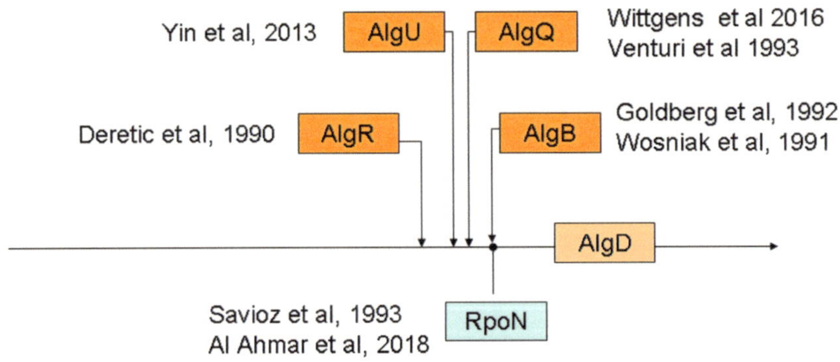

Figure 6.
Regulators of AlgD in alginate production. Negative and positive regulators found up-today.

of glutamine to 2-ketoglutarate, which is dependent on O_2 availability [108, 110]. Another positive regulator of AlgD is AlgU [111], but the only negative regulator known is the RpoN, a sigma factor, that regulates nitrogen metabolism. RpoN is increased by disruption of pyrimidine synthesis and decreased by the supplementation with uracil, showing that a high level of RpoN, in the RSCV form, may block the alginate biosynthesis [110, 112].

Studies on the biosynthetic pathway of biofilms show that chelation of iron by lactoferrin destabilizes the bacterial membrane [113], which combined with xylitol hinders the ability of the bacteria to respond to iron deficiency [101], showing some promise for CF treatment.

4. Conclusions

P. aeruginosa is a relevant pathogen given its widespread prevalence across different organs. The latent menace it poses for inpatients is a liability for institutions. For this, and the negative prognosis that *P. aeruginosa* infections in CF patients has, it is one of the subjects more researched for the last 40 years. The efforts have resulted in understanding the process of invasion, immune response, and bacterial tactics to achieve chronic infection. The complexity of the metabolic changes caused by the contact between the host and the bacteria is so extensive that the selection of variables for in vitro studies is difficult since the production of biofilm by *P. aeruginosa* seems to be regulated by everything, O_2, N_2, Fe^{2+}, Pi, and NaCl. This multiregulatory network is still a puzzle to be resolved.

Scientists agree that suppression of alginate production is vital to treat CF patients, but in 40 years of research, little has been achieved in suppressing its production *in vivo.*

5. Perspectives

The advancement of techniques with high output data like microarrays, proteomes, and mass spectrometry are closing the breach among the different approaches that have been used to tackle *P. aeruginosa* infections. For example, mass-spectrometry has verified through metabolite detection the metabolic pathways studied by molecular biologists and enzymologists. The integration of these studies with the physicians is needed to assess the areas that show more promises to control alginate production and *P. aeruginosa* eradication after it became a chronic infection.

Author details

Sandra Grumelli
Independent Researcher, Centro de Investigaciones en Medicina Respiratoria, Córdoba, Argentina

*Address all correspondence to: sgrumelli@yahoo.com

IntechOpen

© 2019 The Author(s). Licensee IntechOpen. This chapter is distributed under the terms of the Creative Commons Attribution License (http://creativecommons.org/licenses/by/3.0), which permits unrestricted use, distribution, and reproduction in any medium, provided the original work is properly cited.

References

[1] Dean HF, Royle P, Morgan AF. Detection of FP plasmids in hospital isolates of *Pseudomonas aeruginosa*. Journal of Bacteriology. 1979;**138**(1):249-250

[2] Hoadley AW, Knight DE. External otitis among swimmers and nonswimmers. Archives of Environmental Health. 1975;**30**(9):445-448

[3] Crull MR, Ramos KJ, Caldwell E, Mayer-Hamblett N, Aitken ML, Goss CH. Change in *Pseudomonas aeruginosa* prevalence in cystic fibrosis adults over time. BMC Pulmonary Medicine. 2016;**16**(1):176

[4] MacKenzie T, Gifford AH, Sabadosa KA, Quinton HB, Knapp EA, Goss CH, et al. Longevity of patients with cystic fibrosis in 2000 to 2010 and beyond: Survival analysis of the cystic fibrosis foundation patient registry. Annals of Internal Medicine. 2014;**161**(4):233-241. DOI: 10.7326/M13-0636

[5] Knudsen PK, Olesen HV, Høiby N, Johannesson M, Karpati F, Laerum BN, et al. Differences in prevalence and treatment of *Pseudomonas aeruginosa* in cystic fibrosis centres in Denmark, Norway and Sweden. Journal of Cystic Fibrosis. 2009;**8**(2):135-142. DOI: 10.1016/j.jcf.2008.11.001. Epub 2009 Jan 20

[6] Varaiya A, Kulkarni M, Bhalekar P, Dogra J. Incidence of metallo-beta-lactamase-producing *Pseudomonas aeruginosa* in diabetes and cancer patients. Indian Journal of Pathology and Microbiology. 2008;**51**(2):200-203

[7] Trecarichi EM, Tumbarello M. Antimicrobial-resistant Gram-negative bacteria in febrile neutropenic patients with cancer: Current epidemiology and clinical impact. Current Opinion in Infectious Diseases. 2014;**27**(2):200-210

[8] Rolston KV, Bodey GP. *Pseudomonas aeruginosa* infection in cancer patients. Cancer Investigation. 1992;**10**(1):43-59

[9] Corredoira JM, Ariza J, Pallarés R, Carratalá J, Viladrich PF, Rufí G, et al. Gram-negative bacillary cellulitis in patients with hepatic cirrhosis. European Journal of Clinical Microbiology & Infectious Diseases. 1994;**13**(1):19-24

[10] Ormerod LD, Smith RE. Contact lens-associated microbial keratitis. Archives of Ophthalmology. 1986;**104**(1):79-83

[11] Huber-Spitzy V, Baumgartner I, Arocker-Mettinger E, Schiffbänker M, Georgiew L, Grabner G. Corneal ulcer. Current analysis from specialized ambulatory care of a clinic. Klinische Monatsblätter für Augenheilkunde. 1992;**200**(4):251-256

[12] Singh G, Palanisamy M, Madhavan B, Rajaraman R, Narendran K, Kour A, et al. Multivariate analysis of childhood microbial keratitis in South India. Annals of the Academy of Medicine, Singapore. 2006;**35**(3):185-189

[13] Watt KG, Swarbrick HA. Trends in microbial keratitis associated with orthokeratology. Eye & Contact Lens. 2007;**33**(6 Pt 2):373-377; discussion 382

[14] Menon KV, Sorour TM. Epidemiologic and demographic attributes of primary spondylodiscitis in a middle eastern population sample. World Neurosurgery. 2016;**95**:31-39. DOI: 10.1016/j.wneu.2016.07.088. Epub 2016 Aug 2

[15] Lee MK, Chiu CS, Chow VC, Lam RK, Lai RW. Prevalence of hospital infection and antibiotic use at a university medical center in Hong Kong. Journal of Hospital Infection. 2007;**65**(4):341-347. Epub 2007 Feb 2

[16] Boucher HW, Talbot GH, Bradley JS, Edwards JE, Gilbert D, Rice LB, et al. Bad bugs, no drugs: No ESKAPE! An update from the Infectious Diseases Society of America. Clinical Infectious Diseases. 2009;**48**(1):1-12. DOI: 10.1086/595011

[17] Engler K, Mühlemann K, Garzoni C, Pfahler H, Geiser T, von Garnier C. Colonisation with *Pseudomonas aeruginosa* and antibiotic resistance patterns in COPD patients. Swiss Medical Weekly. 2012;**142**:w13509. DOI: 10.4414/smw.2012.13509. eCollection 2012

[18] Gallego M, Pomares X, Espasa M, Castañer E, Solé M, Suárez D, et al. *Pseudomonas aeruginosa* isolates in severe chronic obstructive pulmonary disease: Characterization and risk factors. BMC Pulmonary Medicine. 2014;**14**:103. DOI: 10.1186/1471-2466-14-103

[19] Lieberman D, Lieberman D. Pseudomonal infections in patients with COPD: Epidemiology and management. American Journal of Respiratory Medicine. 2003;**2**(6):459-468

[20] Huerta A, Soler N, Esperatti M, Guerrero M, Menendez R, Gimeno A, et al. Importance of *Aspergillus* spp. isolation in acute exacerbations of severe COPD: Prevalence, factors and follow-up: The FUNGI-COPD study. Respiratory Research. 2014;**15**:17. DOI: 10.1186/1465-9921-15-17

[21] Planquette B, Péron J, Dubuisson E, Roujansky A, Laurent V, Le Monnier A, et al. Antibiotics against *Pseudomonas aeruginosa* for COPD exacerbation in ICU: A 10-year retrospective study. International Journal of Chronic Obstructive Pulmonary Disease. 2015;**10**:379-388. DOI: 10.2147/COPD.S71413. eCollection 2015

[22] Emerson J, Rosenfeld M, McNamara S, Ramsey B, Gibson RL. *Pseudomonas aeruginosa* and other predictors of mortality and morbidity in young children with cystic fibrosis. Pediatric Pulmonology. 2002;**34**:91-100. DOI: 10.1002/ppul.10127

[23] Gibson RL, Burns JL, Ramsey BW. Pathophysiology and management of pulmonary infections in cystic fibrosis. American Journal of Respiratory and Critical Care Medicine. 2003;**168**:918-951. DOI: 10.1164/rccm.200304-505SO

[24] Perez MM, Luna MC, Pivetta OH, Keyeux G. CFTR gene analysis in Latin American CF patients: Heterogeneous origin and distribution of mutations across the continent. Journal of Cystic Fibrosis. 2007;**6**:194-208

[25] Perone C, Medeiros GS, del Castillo DM, de Aguiar MJB, Januário JN. Frequency of 8 CFTR gene mutations in cystic fibrosis patients in Minas Gerais, Brazil, diagnosed by neonatal screening. Brazilian Journal of Medical and Biological Research. 2010;**43**(2):134-138

[26] Drzymała-Czyż S, Szczepanik M, Krzyżanowska P, Duś-Żuchowska M, Pogorzelski A, Sapiejka E, et al. Serum phospholipid fatty acid composition in cystic fibrosis patients with and without liver cirrhosis. Annals of Nutrition & Metabolism. 2017;**71**(1-2):91-98. DOI: 10.1159/000477913. Epub 2017 Jul 22

[27] Chen WC, Huang JW, Chen KY, Hsueh PR, Yang PC. Spontaneous bilateral bacterial empyema in a patient with nephrotic syndrome. The Journal of Infection. 2006;**53**(3):e131-e134. Epub 2006 Feb 7

[28] Fernández J, Acevedo J, Castro M, Garcia O, de Lope CR, Roca D, et al. Prevalence and risk factors of infections by multiresistant bacteria in cirrhosis: A prospective study. Hepatology. 2012;**55**(5):1551-1561. DOI: 10.1002/hep.25532. Epub 2012 Apr 4

[29] Widdicombe J. In: Stockley RA, editor. Pulmonary Defences. Chichester: Wiley; 1997. pp. 1-15

[30] Joris L, Dab I, Quinton PM. Elemental composition of human airway surface fluid in healthy and diseased airways. American Review of Respiratory Disease. 1993;**148**(6 Pt 1):1633-1637

[31] Veldhuizen RAW, Nag K, Orgeig S, Possmayer F. The role of lipids in pulmonary surfactant. Biochimica et Biophysica Acta—Molecular Basis of Disease. 1998;**1408**:90-108

[32] Bufler P, Schmidt B, Schikor D, Bauernfeind A, Crouch EC, Griese M. Surfactant protein A and D differently regulate the immune response to nonmucoid *Pseudomonas aeruginosa* and its lipopolysaccharide. American Journal of Respiratory Cell and Molecular Biology. 2003;**28**(2):249-256

[33] Perez-Gil J, Weaver TE. Pulmonary surfactant pathophysiology: Current models and open questions. Cytokine stimulation by *Pseudomonas aeruginosa*—Strain variation and modulation by pulmonary surfactant. Physiology. 2010;**25**:132-141

[34] Hickling TP, Sim RB, Malhotra R. Induction of TNF-alpha release from human buffy coat cells by *Pseudomonas aeruginosa* is reduced by lung surfactant protein A. FEBS Letters. 1998;**437**(1-2):65-69

[35] Adamo R, Sokol S, Soong G, Gomez M, Prince A. *P. aeruginosa* flagella activate airway epithelial cells through asialo GM1 and TLR2 as well as TLR5. American Journal of Respiratory Cell and Molecular Biology. 2004;**30**:627-634

[36] Zhang Z, Louboutin JP, Weiner DJ, Goldberg JB, Wilson JM. Human airway epithelial cells sense *Pseudomonas aeruginosa* infection via recognition of flagellin by Toll-like receptor 5. Infection and Immunity. 2005;**73**:7151-7160

[37] Ramphal R, Balloy V, Huerre M, Si-Tahar M, Chignard M. TLRs 2 and 4 are not involved in hypersusceptibility to acute *Pseudomonas aeruginosa* lung infections. Journal of Immunology. 2005;**175**:3927-3934

[38] Power MR, Peng Y, Maydanski E, Marshall JS, Lin TJ. The development of early host response to *Pseudomonas aeruginosa* lung infection is critically dependent on myeloid differentiation factor 88 in mice. The Journal of Biological Chemistry. 2004;**279**:49315-49322

[39] Flo TH, Ryan L, Latz E, Takeuchi O, Monks BG, Lien E, et al. Involvement of Toll-like receptor (TLR) 2 and TLR4 in cell activation by mannuronic acid polymers. The Journal of Biological Chemistry. 2002;**277**:35489-35495

[40] Epelman S, Stack D, Bell C, Wong E, Neely GG, Krutzik S, et al. Different domains of *Pseudomonas aeruginosa* exoenzyme S activate distinct TLRs. Journal of Immunology. 2004;**173**:2031-2040

[41] Parker D, Prince A. Epithelial uptake of flagella initiates proinflammatory signaling. PLoS One. 2013;**8**(3):e59932. DOI: 10.1371/journal.pone.0059932. Epub 2013 Mar 20

[42] Schroeder TH, Lee MM, Yacono PW, Cannon CL, Gerçeker AA, Golan DE, et al. CFTR is a pattern recognition molecule that extracts *Pseudomonas aeruginosa* LPS from the outer membrane into epithelial cells and activates NF-kappa B translocation. Proceedings of the National Academy of Sciences of the United States of America. 2002;**99**(10):6907-6912

[43] Ichikawa JK, Norris A, Bangera MG, Geiss GK, van 't Wout AB, Bumgarner RE, et al. Interaction of *Pseudomonas aeruginosa* with epithelial cells: Identification of differentially regulated genes by expression microarray analysis of human cDNAs. Proceedings of the National Academy of Sciences

of the United States of America. 2000;**97**(17):9659-9664

[44] Reiniger N, Ichikawa JK, Pier GB. Influence of cystic fibrosis transmembrane conductance regulator on gene expression in response to *Pseudomonas aeruginosa* infection of human bronchial epithelial cells. Infection and Immunity. 2005;**73**(10):6822-6830

[45] Joseph T, Look D, Ferkol T. NF-kappaB activation and sustained IL-8 gene expression in primary cultures of cystic fibrosis airway epithelial cells stimulated with *Pseudomonas aeruginosa*. American Journal of Physiology. Lung Cellular and Molecular Physiology. 2005;**288**(3):L471-L479

[46] Ribeiro CM, Paradiso AM, Schwab U, Perez-Vilar J, Jones L, O'neal W, et al. Chronic airway infection/inflammation induces a Ca^{2+}i-dependent hyperinflammatory response in human cystic fibrosis airway epithelia. The Journal of Biological Chemistry. 2005;**280**(18):17798-17806

[47] Véliz Rodriguez T, Moalli F, Polentarutti N, Paroni M, Bonavita E, Anselmo A, et al. Role of Toll interleukin-1 receptor (IL-1R) 8, a negative regulator of IL-1R/Toll-like receptor signaling, in resistance to acute *Pseudomonas aeruginosa* lung infection. Infection and Immunity. 2012;**80**(1):100-109. DOI: 10.1128/IAI.05695-11. Epub 2011 Oct 24

[48] Björkbacka H, Fitzgerald KA, Huet F, Li X, Gregory JA, Lee MA, et al. The induction of macrophage gene expression by LPS predominantly utilizes Myd88-independent signaling cascades. Physiological Genomics. 2004;**19**:319-330

[49] Weighardt H, Kaiser-Moore S, Vabulas RM, Kirschning CJ, Wagner H, Holzmann B. Cutting edge: Myeloid differentiation factor 88 deficiency improves resistance against sepsis caused by polymicrobial infection. Journal of Immunology. 2002;**169**:2823-2827

[50] Shi S, Nathan C, Schnappinger D, Drenkow J, Fuortes M, Block E, et al. MyD88 primes macrophages for full-scale activation by interferon-γ yet mediates few responses to *Mycobacterium tuberculosis*. The Journal of Experimental Medicine. 2003;**198**:987-997

[51] Skerrett SJ, Liggitt HD, Hajjar AM, Wilson CB. Cutting edge: Myeloid differentiation factor 88 is essential for pulmonary host defense against *Pseudomonas aeruginosa* but not *Staphylococcus aureus*. Journal of Immunology. 2004;**172**:3377-3381

[52] Liang X, Gupta K, Quintero JR, Cernadas M, Kobzik L, Christou H, et al. Macrophage FABP4 is required for neutrophil recruitment and bacterial clearance in *Pseudomonas aeruginosa* pneumonia. FASEB Journal. 2018:fj201802002R. DOI: 10.1096/fj.201802002R

[53] Tiesset H, Pierre M, Desseyn JL, Guéry B, Beermann C, Galabert C, et al. Dietary (n-3) polyunsaturated fatty acids affect the kinetics of pro- and antiinflammatory responses in mice with *Pseudomonas aeruginosa* lung infection. The Journal of Nutrition. 2009;**139**(1):82-89. DOI: 10.3945/jn.108.096115. Epub 2008 Dec 3

[54] Bayes HK, Ritchie ND, Evans TJ. Interleukin-17 is required for control of chronic lung infection caused by *Pseudomonas aeruginosa*. Infection and Immunity. 2016;**84**(12):3507-3516

[55] Ding FM, Zhang XY, Chen YQ, Liao RM, Xie GG, Zhang PY, et al. Lentivirus-mediated overexpression of suppressor of cytokine signaling-3 reduces neutrophilic airway inflammation by suppressing T-helper 17 responses in mice with chronic *Pseudomonas*

aeruginosa lung infections. International Journal of Molecular Medicine. 2018;**41**(4):2193-2200. DOI: 10.3892/ijmm.2018.3417. Epub 2018 Jan 23

[56] Pan T, Tan R, Li M, Liu Z, Wang X, Tian L, et al. IL17-producing γδ T cells may enhance humoral immunity during pulmonary *Pseudomonas aeruginosa* infection in mice. Frontiers in Cellular and Infection Microbiology. 2016;**6**:170. DOI: 10.3389/fcimb.2016.00170. eCollection 2016

[57] Li JL, Chen TS, Yuan CC, Zhao GQ, Xu M, Li XY, et al. Regulatory T cell activity in immunosuppresive mice model of *Pseudomonas aeruginosa pneumonia*. Journal of Huazhong University of Science and Technology. Medical Sciences. 2017;**37**(4):505-509. DOI: 10.1007/s11596-017-1764-2. Epub 2017 Aug 8

[58] Hu ZQ, Yao YM, Chen W, Bian JL, Zhao LJ, Chen LW, et al. Partial depletion of regulatory T cells enhances host inflammatory response against acute *Pseudomonas aeruginosa* infection after sepsis. Inflammation. 2018;**41**(5):1780-1790. DOI: 10.1007/s10753-018-0821-8

[59] Shindo Y, Fuchs AG, Davis CG, Eitas T, Unsinger J, Burnham CD, et al. Interleukin 7 immunotherapy improves host immunity and survival in a two-hit model of *Pseudomonas aeruginosa* pneumonia. Journal of Leukocyte Biology. 2017;**101**(2):543-554. DOI: 10.1189/jlb.4A1215-581R. Epub 2016 Sep 14

[60] Chen W, Lian J, Ye JJ, Mo QF, Qin J, Hong GL, et al. Ethyl pyruvate reverses development of *Pseudomonas aeruginosa* pneumonia during sepsis-induced immunosuppression. International Immunopharmacology. 2017;**52**:61-69. DOI: 10.1016/j.intimp.2017.08.024. Epub 2017 Aug 31

[61] Pugh AM, Auteri NJ, Goetzman HS, Caldwell CC, Nomellini V. A murine model of persistent inflammation, immune suppression, and catabolism syndrome. International Journal of Molecular Sciences. 2017;**18**(8);pii: E1741. DOI: 10.3390/ijms18081741

[62] Naughton S, Parker D, Seemann T, Thomas T, Turnbull L, Rose B, et al. *Pseudomonas aeruginosa* AES-1 exhibits increased virulence gene expression during chronic infection of cystic fibrosis lung. PLoS One. 2011;**6**(9):e24526. DOI: 10.1371/journal.pone.0024526. Epub 2011 Sep 15

[63] Bricio-Moreno L, Sheridan VH, Goodhead I, Armstrong S, Wong JKL, Waters EM, et al. Evolutionary trade-offs associated with loss of PmrB function in host-adapted *Pseudomonas aeruginosa*. Nature Communications. 2018;**9**(1):2635. DOI: 10.1038/s41467-018-04996-x

[64] Grumelli S. Choline triggers exacerbations of chronic obstructive pulmonary disease in patients infected with *Pseudomonas aeruginosa*. Current Respiratory Medicine Reviews. 2016;**12**(2):167-174. DOI: 10.2174/1573398X12999160506104327

[65] Déziel E, Comeau Y, Villemur R. Initiation of biofilm formation by *Pseudomonas aeruginosa* 57RP correlates with emergence of hyperpiliated and highly adherent phenotypic variants deficient in swimming, swarming, and twitching motilities. Journal of Bacteriology. 2001;**183**:1195-1204

[66] Drenkard E, Ausubel FM. Pseudomonas biofilm formation and antibiotic resistance are linked to phenotypic variation. Nature. 2002;**416**:740-743

[67] von Götz F, Häussler S, Jordan D, Saravanamuthu SS, Wehmhöner D, Strüssmann A, et al. Expression analysis of a highly adherent and cytotoxic small colony variant of *Pseudomonas aeruginosa* isolated from a lung of a

patient with cystic fibrosis. Journal of Bacteriology. 2004;**186**(12):3837-3847

[68] Pu M, Sheng L, Song S, Gong T, Wood TK. Serine hydroxymethyltransferase ShrA (PA2444) controls rugose small-colony variant formation in *Pseudomonas aeruginosa*. Frontiers in Microbiology. 2018;**9**:315. DOI: 10.3389/fmicb.2018.00315. eCollection 2018

[69] Rao NN, Liu S, Kornberg A. Inorganic polyphosphate in *Escherichia coli*: The phosphate regulon and the stringent response. Journal of Bacteriology. 1998;**180**(8):2186-2193

[70] Malone JG, Jaeger T, Spangler C, Ritz D, Spang A, Arrieumerlou C, et al. YfiBNR mediates cyclic di-GMP dependent small colony variant formation and persistence in *Pseudomonas aeruginosa*. PLoS Pathogens. 2010;**6**(3):e1000804. DOI: 10.1371/journal.ppat.1000804

[71] Evans TJ. Small colony variants of *Pseudomonas aeruginosa* in chronic bacterial infection of the lung in cystic fibrosis. Future Microbiology. 2015;**10**(2):231-239. DOI: 10.2217/fmb.14.107

[72] Malone JG, Jaeger T, Manfredi P, Dötsch A, Blanka A, Bos R, et al. The YfiBNR signal transduction mechanism reveals novel targets for the evolution of persistent *Pseudomonas aeruginosa* in cystic fibrosis airways. PLOS Pathogens. 2012;**8**(6):e1002760. DOI: 10.1371/journal.ppat.1002760. Epub 2012 Jun 14

[73] Grumelli SM. Bioenergetics of phosphate incorporation and polyphosphate accumulation by choline in Pseudomonas aeruginosa [doctoral thesis]. 2000. Ejemplares 59118

[74] Lim WS, Phang KK, Tan AH, Li SF, Ow DS. Small colony variants and single nucleotide variations in Pf1 region of PB1 phage-resistant *Pseudomonas aeruginosa*. Frontiers in Microbiology. 2016;**7**:282. DOI: 10.3389/fmicb.2016.00282. eCollection 2016

[75] Nikel PI, Chavarría M, Martínez-García E, Taylor AC, de Lorenzo V. Accumulation of inorganic polyphosphate enables stress endurance and catalytic vigour in *Pseudomonas putida* KT2440. Microbial Cell Factories. 2013;**12**:50. DOI: 10.1186/1475-2859-12-50

[76] Landry RM, An D, Hupp JT, Singh PK, Parsek MR. Mucin-*Pseudomonas aeruginosa* interactions promote biofilm formation and antibiotic resistance. Molecular Microbiology. 2006;**59**(1):142-151

[77] Zdorovenko EL, Zatonskii GV, Kocharova NA, Shashkov AS, Knirel YA, Ovod VV. Structure of the O polysaccharides and serological classification of *Pseudomonas syringae* pv. *porri* from genomospecies 4. European Journal of Biochemistry. 2003;**270**(1):20-27

[78] Bystrova OV, Shashkov AS, Kocharova NA, Knirel YA, Zähringer U, Pier GB. Elucidation of the structure of the lipopolysaccharide core and the linkage between the core and the O-antigen in *Pseudomonas aeruginosa* immunotype 5 using strong alkaline degradation of the lipopolysaccharide. Biochemistry (Mosc). 2003;**68**(8):918-925

[79] Olvera C, Goldberg JB, Sánchez R, Soberón-Chávez G. The *Pseudomonas aeruginosa* algC gene product participates in rhamnolipid biosynthesis. FEMS Microbiology Letters. 1999;**179**(1):85-90

[80] Wargo MJ, Gross MJ, Rajamani S, et al. Hemolytic phospholipase c inhibition protects lung function during *Pseudomonas aeruginosa*. American Journal of Respiratory and Critical Care Medicine. 2011;**184**:345-354

[81] Salvano MA, Domenech CE. Kinetic properties of purified *Pseudomonas aeruginosa* phosphorylcholine phosphatase indicated that this enzyme may be utilized by the bacteria to colonize in different environments. Current Microbiology. 1999;**39**:1-8

[82] Kadurugamuwa JL, Beveridge TJ. Virulence factors are released from *Pseudomonas aeruginosa* in association with membrane vesicles during normal growth and exposure to gentamicin: A novel mechanism of enzyme secretion. Journal of Bacteriology. 1995;**177**(14):3998-4008

[83] Fuchs B, Rupp M, Ghofrani HA, et al. Diacylglycerol regulates acute hypoxic pulmonary vasoconstriction via TRPC6. Respiratory Research. 2011;**12**:20

[84] Drake J, Glavinović MI, Trifaro JM. Choline blockage of currents through Ca^{2+} activated K^{+} channels in bovine chromaffin cells. Neuroscience. 1992;**49**:945-950

[85] LeGouëllec A, Moyne O, Meynet E, Toussaint B, Fauvelle F. High-resolution magic angle spinning NMR-based metabolomics revealing metabolic changes in lung of mice infected with *P. aeruginosa* consistent with the degree of disease severity. Journal of Proteome Research. 2018;**17**(10):3409-3417. DOI: 10.1021/acs.jproteome.8b00306. Epub 2018 Sep 14

[86] Salvano MA, Lisa TA, Domenech CE. Choline transport in *Pseudomonas aeruginosa*. Molecular and Cellular Biochemistry. 1989;**85**:81-89

[87] Malek AA, Chen C, Wargo MJ, Beattie GA, Hogan DA. Roles of three transporters, CbcXWV, BetT1, and BetT3, in *Pseudomonas aeruginosa* choline uptake for catabolism. Journal of Bacteriology. 2011;**193**(12):3033-3041

[88] Sage AE, Vasil AI, Vasil ML. Molecular characterization of mutants affected in the osmoprotectant-dependent induction of phospholipase C in *Pseudomonas aeruginosa* PAO1. Molecular Microbiology. 1997;**23**(1):43-56

[89] Diab F, Bernard T, Bazire A, Haras D, Blanco C, Jebbar M. Succinate-mediated catabolite repression control on the production of glycine betaine catabolic enzymes in *Pseudomonas aeruginosa* PAO1 under low and elevated salinities. Microbiology. 2006;**152**(Pt 5):1395-1406

[90] Wargo MJ, Szwergold BS, Hogan DA. Identification of two gene clusters and a transcriptional regulator required for *Pseudomonas aeruginosa* glycine betaine catabolism. Journal of Bacteriology. 2008;**190**(8):2690-2699. Epub 2007 Oct 19

[91] Wargo MJ. Choline catabolism to glycine betaine contributes to *Pseudomonas aeruginosa* survival during murine lung infection. PLoS One. 2013;**8**(2):e56850. DOI: 10.1371/journal.pone.0056850. Epub 2013 Feb 14

[92] Le Rudulier D, Strøm AR, Dandekar AM, Smith LT, Valentine RC. Molecular biology of osmoregulation. Science. 1984;**224**:1064-1068

[93] Lisa TA, Garrido MN, Domenech CE. Induction of acid phosphatase and cholinesterase activities in *Ps. aeruginosa* and their in-vitro control by choline, acetylcholine and betaine. Molecular and Cellular Biochemistry. 1983;**50**:149-155

[94] Bray C, Son JH, Kumar P, Harris JD, Meizel S. A role for the human sperm glycine receptor/Cl(−) channel in the acrosome reaction initiated by recombinant ZP3. Biology of Reproduction. 2002;**66**(1):91-97

[95] Wheeler MD, Thurman RG. Production of superoxide and TNF-alpha from alveolar macrophages is

blunted by glycine. American Journal of Physiology. 1999;**277**(5):L952-L959. DOI: 10.1152/ajplung.1999.277.5.L952

[96] Pavliakova D, Chu C, Bystricky S, et al. Treatment with succinic anhydride improves the immunogenicity of *Shigella flexneri* type 2a O-specific polysaccharide-protein conjugates in mice. Infection and Immunity. 1999;**67**(10):5526-5529

[97] Greer GG, Milazzo FH. *Pseudomonas aeruginosa* lipopolysaccharide: An uncoupler of mitochondrial oxidative phosphorylation. Canadian Journal of Microbiology. 1975;**21**(6):877-883

[98] Joo SH, Chung HS, Raetz CR, Garrett TA. Activity and crystal structure of *Arabidopsis thaliana* UDP-N-acetylglucosamine acyltransferase. Biochemistry. 2012;**51**(21):4322-4330. DOI: 10.1021/bi3002242. Epub 2012 May 14

[99] Al-Hasan MN, Juhn YJ, Bang DW, Yang HJ, Baddour LM. External validation of bloodstream infection mortality risk score in a population-based cohort. Clinical Microbiology and Infection. 2014;**20**(9):886-891. DOI: 10.1111/1469-0691.12607. Epub 2014 Mar 26

[100] Litwin CM, Calderwood SB. Role of iron in regulation of virulence genes. Clinical Microbiology Reviews. 1993;**6**:137-149

[101] Wittgens A, Kovacic F, Müller MM, Gerlitzki M, Santiago-Schübel B, Hofmann D, et al. Novel insights into biosynthesis and uptake of rhamnolipids and their precursors. Applied Microbiology and Biotechnology. 2017;**101**(7):2865-2878. DOI: 10.1007/s00253-016-8041-3. Epub 2016 Dec 17

[102] Venturi V, Ottevanger C, Leong J, Weisbeek PJ. Identification and characterization of a siderophore regulatory gene (pfrA) of *Pseudomonas putida* WCS358: Homology to the alginate regulatory gene algQ of *Pseudomonas aeruginosa*. Molecular Microbiology. 1993;**10**(1):63-73

[103] Bereswill S, Waidner U, Odenbreit S, Lichte F, Fassbinder F, Bode G, et al. Structural, functional and mutational analysis of the pfr gene encoding a ferritin from *Helicobacter pylori*. Microbiology. 1998;**144**(Pt 9):2505-2516

[104] Linker A, Jones RS. A new polysaccharide resembling alginic acid isolated from pseudomonads. Journal of Biological Chemistry. 1966;**241**:3845-3851

[105] Deretic V, Gill JF, Chakrabarty AM. Gene algD coding for GDPmannose dehydrogenase is transcriptionally activated in mucoid *Pseudomonas aeruginosa*. Journal of Bacteriology. 1987;**169**(1):351-358

[106] Cunin R, Glansdorff N, Pierard A, Stalon V. Biosynthesis and metabolism of arginine in bacteria. Microbiological Review. 1986;**50**:314-352

[107] Park SM, Lu CD, Abdelal AT. Cloning and characterization of argR, a gene that participates in regulation of arginine biosynthesis and catabolism in *Pseudomonas aeruginosa* PAO1. Journal of Bacteriology. 1997;**179**(17):5300-5308

[108] Deretic V, Govan JR, Konyecsni WM, Martin DW. Mucoid *Pseudomonas aeruginosa* in cystic fibrosis: Mutations in the muc loci affect transcription of the algR and algD genes in response to environmental stimuli. Molecular Microbiology. 1990;**4**(2):189-196

[109] Wang Y, Hay ID, Rehman ZU, Rehm BH. Membrane-anchored MucR mediates nitrate-dependent regulation of alginate production in *Pseudomonas aeruginosa*. Applied Microbiology and Biotechnology. 2015;**99**(17):7253-7265. DOI: 10.1007/s00253-015-6591-4. Epub 2015 Apr 29

[110] Savioz A, Zimmermann A, Haas D. *Pseudomonas aeruginosa* promoters which contain a conserved GG-N10-GC motif but appear to be RpoN-independent. Molecular Genetics and Genomics. 1993;**238**(1-2):74-80

[111] Yin Y, Withers TR, Wang X, Yu HD. Evidence for sigma factor competition in the regulation of alginate production by *Pseudomonas aeruginosa*. PLoS One. 2013;**8**(8):e72329. DOI: 10.1371/journal.pone.0072329. eCollection 2013

[112] Al Ahmar R, Kirby BD, Yu HD. Pyrimidine biosynthesis regulates small colony variant and mucoidy in *Pseudomonas aeruginosa* through sigma factor competition. Journal of Bacteriology. 2018. pii: JB.00575-18. DOI: 10.1128/JB.00575-18

[113] Ammons CM, Ward LS, Fisher ST, Wolcott RD, James GA. In vitro susceptibility of established biofilms composed of a clinical wound isolate of *Pseudomonas aeruginosa* treated with lactoferrin and xylitol. International Journal of Antimicrobial Agents. 2009;**33**(3):230-236

Chapter 4

The Role of *Pseudomonas aeruginosa* RNA Methyltransferases in Antibiotic Resistance

Pablo Valderrama-Carmona, Jaison H. Cuartas, Diana Carolina Castaño and Mauricio Corredor

Abstract

Methyltransferases play a fundamental role in aminoglycoside resistance of Gram-negative bacteria, and some of its mechanisms were described in the past years, especially in *Escherichia coli*; however, it remains unsolved for other resistant bacteria such as *Pseudomonas aeruginosa*. Despite hurdles to determine resistance acquisition, high-throughput approaches (genomics, transcriptomics, and proteomics) have allowed data mining and analysis in a systemic way. Likewise, bioinformatics modelling of homologous genes or proteins has permitted to elucidate the emerging resistance in this pathogen. *P. aeruginosa* is a bacterial resistance treat since practically all known resistance mechanisms can be described using this model, particularly RNA methyltransferases. The RNA methyltransferases perform methylation or demethylation of ribosomal RNA to allow or restrict the antibiotic resistance development. The Kgm and Kam methyltransferases families are found in P. aeruginosa and confer resistance to several aminoglycosides. Loss of native methylations may also confer a resistant phenotype. The P. aeruginosa RsmG has high structural homology with Thermus aquaticus protein. Today, molecular data will promote a new paradigm on antibiotic therapy for treatment against *P. aeruginosa*. This chapter provides an overview of what role *P. aeruginosa*'s methyltransferases play in antibiotic resistance, induced by methylation or demethylation in the ribosome.

Keywords: *Pseudomonas aeruginosa*, antibiotic resistance, aminoglycoside, methyltransferase, methylation, demethylation, 16S RNA

1. Introduction

Nowadays, aminoglycoside antibiotic regimen remains as a prevailing therapy for the treatment of *Pseudomonas aeruginosa* pathogen, predominantly for respiratory complications in cystic fibrosis patients. However, *P. aeruginosa* strains are emergent multidrug-resistant. Bacteria develop resistance to aminoglycosides by producing aminoglycoside-modifying enzymes such as acetyltransferase, phosphorylase, and adenyltransferase [1]. The bacterial ribosome is a primary antibiotic

IntechOpen

target, but bacteria can acquire resistance by modification of drug-binding sites. More than 50 years of studies in *Escherichia coli* have shown that 16S and 23S rRNAs have methylated nucleotides (**Figure 1**). These molecular modifications are performed by methyltransferases (MTases), which take in charge the transfer of a methyl group from a methyl donor S-adenosyl-l-methionine, better known as AdoMet or SAM [2]. These RNA MTases are diverse in posttranscriptional RNA modification, where single RNA nucleosides are chemically transformed. SAM-dependent MTases are involved in biosynthesis, signal transduction, protein repair, chromatin regulation, and gene silencing [3]. More recently, it was shown that aminoglycoside resistance in *E. coli* has its primary target within the decoding

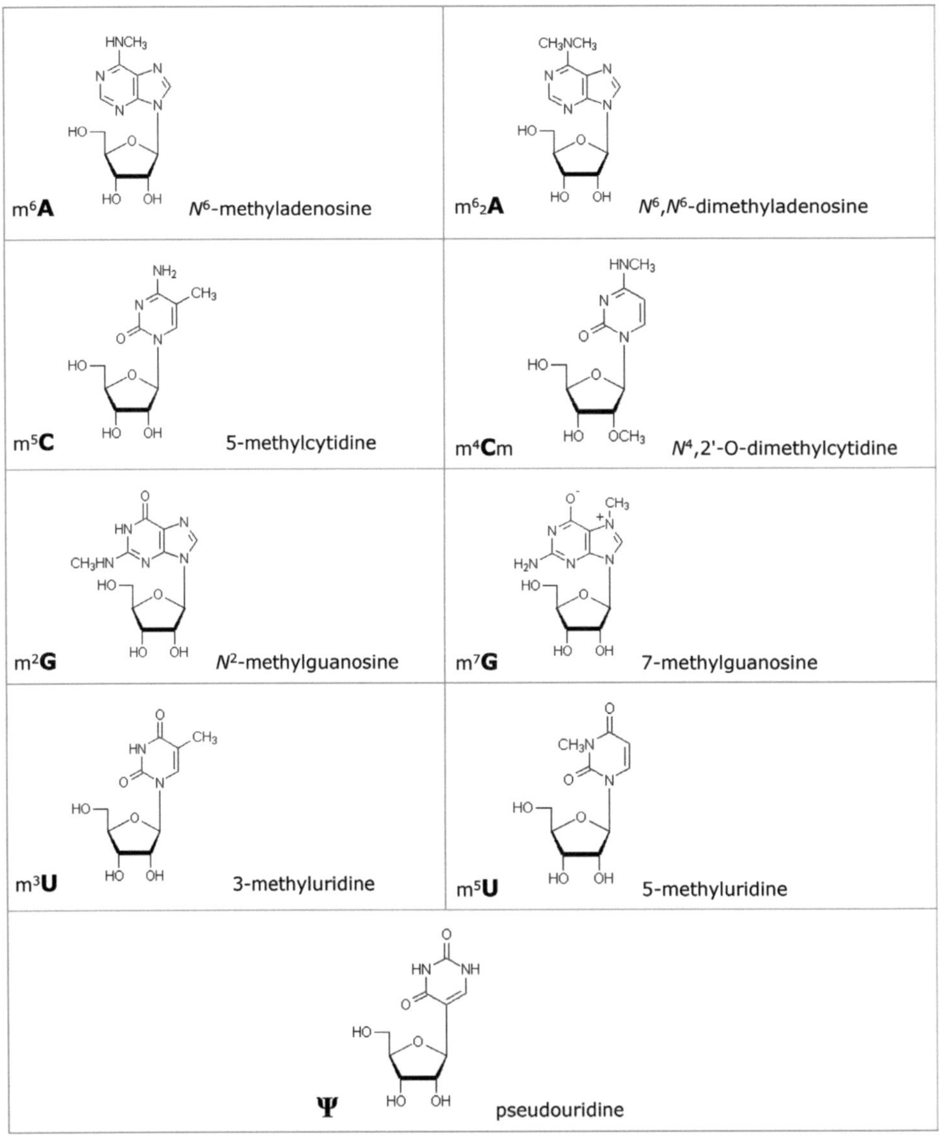

Figure 1.
Ribonucleotides methylated where the methyl moiety is located either in 16S or 23S ribonucleotides: m^6A [66, 67], m^6_2A [68], m^5C [67], m^4Cm [69], m^2G [66], m^7G [70], m^3U [67, 71], m^5U [72], and Ψ [68]. Molecules are designed using the figures on the next web page, https://mods.rna.albany.edu/mods/modifications/search/. The structure shows clearly the methyl (CH_3-) but in the last structure bottom does not show this methyl.

region of 16S and 23S rRNAs and it is known to have 10 methylated nucleosides in 16S rRNA and 14 methylated nucleotides in 23S rRNA [4].

Different methylation sites have been identified within the 16S rRNA which yield different aminoglycoside resistance phenotypes [5]. One type group of 16S rRNA methylases is produced by istamycin producer *Streptomyces tenjimariensis*, which methylates m^1A1408 residue. Another group of 16S rRNA methylases is synthesized by gentamicin producer *Micromonospora purpurea* that methylates residue G1405; nonetheless, in *P. aeruginosa* the modification takes place in helix 44, with a secondary target in 23S rRNA helix 69 [6]. They bind specifically to the aminoacyl site (A-site) of 16S rRNA within the prokaryotic 30S ribosomal subunits and interfere with protein synthesis [7].

Probably RNA methylation began prior to DNA methylation in the early forms of life evolved on Earth [8], allowing to hypothesize perhaps that methylases appeared before polymerases. Ribonucleotide can be methylated by methyltransferases or demethylated by demethyltransferases. The structure of the bacterial ribosome has a molecular mass of 2.5 mega Daltons. In *E. coli*, the 50S subunit is composed of 23S rRNA (2904 nt), 5S rRNA (120 nt), and 33 ribosomal proteins, while the 30S subunit is composed of 16S rRNA (1542 nt) and 21 ribosomal proteins (S1–S21) [9].

The 16S rRNA resistance methyltransferases modify only intact 30S subunits, but the molecular details of their target recognition mechanisms are not quite elucidated yet. Such studies are becoming all the more necessary [10]. RNA methylation has been observed in different types of RNA species, viz., mRNA, rRNA, tRNA, snoRNA, snRNA, miRNA, and tmRNAs. Specific RNA methyltransferases are synthesized by cells to label these RNA species according to their needs and prevailing environmental conditions surrounding the cells, and this molecular labeling system is a constituent of epigenetics. New molecular structures provide crucial new insights that may provide a starting point for strategies to suppress these emerging causes of pathogenic bacterial resistance to aminoglycosides [11]. Nonetheless, bacteria develop resistance to aminoglycosides by producing aminoglycoside-modifying enzymes such as acetyltransferase, phosphorylase, and adenyltransferase. These enzymes, however, cannot confer a broad aminoglycoside resistance spectrum due to its substrate specificity [1].

2. Antibiotic resistance

P. aeruginosa has a large genome among *Gamma proteobacteria*, which allows it to improve many resistance mechanisms in a versatile way, for example, by transmissible plasmids or integrons. *P. aeruginosa* derepresses the chromosomal AmpC cephalosporinase [12, 13]; it also acquires genes for AmpC enzymes, class A carbenicillinases or β-lactamases, class D oxacillinases, and class B carbapenem-hydrolyzing enzymes [14], as it occurs in other bacteria like *E. coli* and *K. pneumoniae*. Other mechanisms include modifying the structure of topoisomerases II and IV to become quinolone resistant [15], decreasing outer membrane permeability by the partial or total failure of OprD proteins [12], overexpressing the active efflux systems with broad substrate patterns [16, 17], or synthesizing aminoglycoside-modifying enzymes as adenylyltransferases, acetyltransferases, and phosphoryltransferases [18].

The range of antibiotic resistance in *P. aeruginosa* is wide, and it represents a major difficulty for health care by its unsuccessful treatment, as a consequence of its low intrinsic antibiotic susceptibility, an effect of the interaction between multidrug efflux pumps, like *mexAB*, *mexXY* [19], *AdeABC*, and *AdeDE* genes [17]. Another factor is its efficient capability of acquiring resistance, developed by transfer of horizontal genes, such as specific gene mutations [20], and finally by the low permeability of the cellular membrane [16, 21].

Protein	Gene	Substrate	Nucleotide methylated	Position of methylation	Ligand	UniProt
RlmB	*rlmB*	23S rRNA	G2251	2'-O-ribose	SAM	Q9HUM8
RlmD	*rlmD*	23S rRNA	U1939	C5	SAM	Q9I525
RlmE	*rlmE*	23S rRNA	U2552	2'-O-ribose	SAM	A6VCK9
RlmF	*rlmF*	23S rRNA	A1618	N6	SAM	A6V0S3
RlmG	*rlmG*	23S rRNA	G1835	N2	SAM	A6VC08
RlmH	*rlmH*	23S rRNA	Ψ1915	N3	SAM	A6V0A6
RlmJ	*rlmJ*	23S rRNA	A2030	N6	SAM	Q9HUF0
RlmK/L	*rlmL*	23S rRNA	G2445	N2	SAM	A6V328
RlmM	*rlmM*	23S rRNA	C2498	2'-O-ribose	SAM	A6V7T6
RlmN	*rlmN*	23S rRNA and tRNA	A2503 in rRNA and A37 in tRNA	C2	Radical SAM	A6V0V7
RsmA	*rsmA*	16S rRNA	A1518 and A1519	N6	SAM	Q9I5U5
RsmB	*rsmB*	16S rRNA	C967	C5	SAM	Q9I7A9
RsmC	*rsmC*	16S rRNA	G1207	N2	SAM	A6VC20
RsmD	*rsmD_2*	16S rRNA	G966	N2	SAM	A0A0F6U8H1
RsmE	*rsmE_2*	16S rRNA	U1498	N3	SAM	A0A0F6U8B3
RsmG	*rsmG*	16S rRNA	G527	N7	SAM	A6VF42
RsmH	*rsmH*	16S rRNA	C1402	N4	SAM	A6VB93
RsmI	*rsmI*	16S rRNA	C1402	2'-O-ribose	SAM	Q9HVZ3
RsmJ	*rsmJ*	16S rRNA	G1516	N2	SAM	Q9HXW0
TrmA	*trmA*	tm/tRNA	U54 in tRNA and U341 in tmRNA	C5	SAM	A6VCH5
TrmB	*trmB*	tRNA	G46	N7	SAM	Q9I6B3
TrmD	*trmD*	tRNA	G37	N1	SAM	Q9HXQ1
TrmH	*trmH*	tRNA(Leu)	Wobble nucleotide	2'-O-ribose	SAM	A0A0H2ZHL8
TrmI	*trmI*	tRNA	A58	N1	SAM	A0A2X4FJT8
TrmJ	*trmJ*	tRNA	C32, U32, and A32	2'-O-ribose	SAM	A0A0H2ZF87
TrmL	*trmL_2*	tRNA(Leu)	Wobble nucleotide	2'-O-ribose	SAM	A0A0G5X8M9

Table 1.
P. aeruginosa's rRNA methyltransferases and their point of modifications [23]. The columns are described as follows: first column, the name protein of RNA methyltransferase; second column, the name of its gene; third column, the substrate either 23S or 16S RNA or tRNA; fourth column, the type of nucleotide methylated; fifth column, the electron in nucleotide methylated; sixth column, the ligand for everyone SAM; and seventh column, the UniProt code. Some interesting proteins such as RsmA, RsmG, RsmH, and RsmI are marked in bold.

Bacterial multidrug resistance (MDR) is an important concern in *P. aeruginosa* since this microorganism is capable of mixing several mechanisms, transposons, plasmids, and chromosomally encoded genes, such as methyltransferases or pumps [22]. Methyltransferase genes are spread in bacterial genome ready to trigger antibiotic resistance. **Table 1** compares the reported methyltransferase proteins worldwide, being annotated for *P. aeruginosa* in UniProt database [23].

One mechanism of adaptation which facilitates natural selection in bacteria is the hypermutation of some genes or chromosomal regions. Previous work in patients with *P. aeruginosa* showed that hypermutation causes a problematic effect during a chronic respiratory infection (CRI) [24], where *P. aeruginosa* was up to 6.5-fold higher in mutator backgrounds. Other elements associated with high antimicrobial resistance are integrons. These elements were found among isolates from Iran patients with *P. aeruginosa*, which the *int1* integron was prevalent [25].

Integrons linked to transposons, plasmids, and chromosome are responsible for bacterial antibiotic resistance [26, 27]. Integrons are composed of three elements: (1) the integrin-associated promoter (Pc), which is required for transcription and expression of gene cassettes (genetic elements that encode antibiotic resistance genes) within the integrin; (2) the *intI* gene, in which coding for the integrase IntI is crucial for site-specific recombination; and (3) the adjacent recombination site *attI*, which is recognized by integrase. On the other hand, *P. aeruginosa* carrying transposon Tn1696 is an element that encodes the *CmlA* gene, an exporter of the major facilitator (MF) superfamily which provides antibiotic resistance, specifically against chloramphenicol [28]. *P. aeruginosa* has a broad spectrum in cephalosporin resistance mechanism, mediated by the extended-spectrum β-lactamases (ESBLs). High prevalence of multidrug resistance in burn patients and production of *oxa-10*, *per-1*, and *veb-1* genes by *P. aeruginosa* isolates confirm the presence of antibiotic-degrading enzymes [29].

The Pathosystems Resource Integration Center (PATRIC) is a massive database that integrates genomic data and analysis tools to support biomedical research on bacterial infectious diseases. The platform provides an interface for biologists to discover data and information and conduct comprehensive comparative genomics and other analyses in a one-step source. PATRIC database provides complete genome information and data regarding susceptibility or resistance [30] to several antibiotics; including aminoglycosides, polymyxin B, colistin, ceftazidime, piperacillin, imipenem, ciprofloxacin, levofloxacin, and meropenem in *P. aeruginosa*. We report

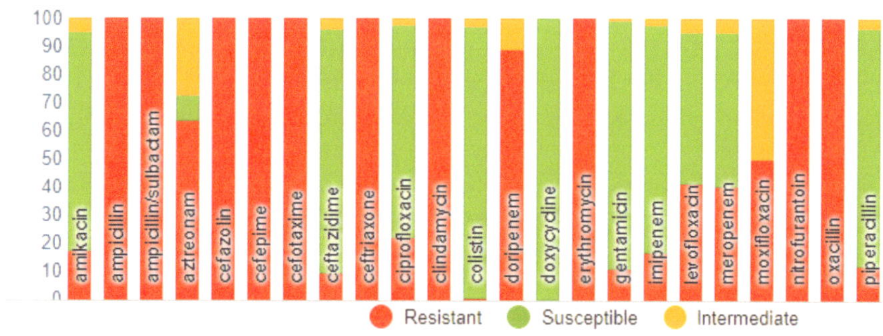

Figure 2.
Resistance and susceptibility profile of P. aeruginosa against a broad spectrum of different types of antibiotics. The data were downloaded from PATRIC database selecting aminoglycosides, beta-lactamases, cephalosporins, licosamides, fluoroquinolones, colistin, doxycycline, ciprofloxacin, nitrofurantoin, and cefazolin. Many strains are resistant to a wide range of antibiotics (red with a larger percentage), and the most strains are susceptible to colistin (green with larger percentage); on the other hand, the overall strains are resistant to ampicillin, cefotaxime, erythromycin, and nitrofurantoin.

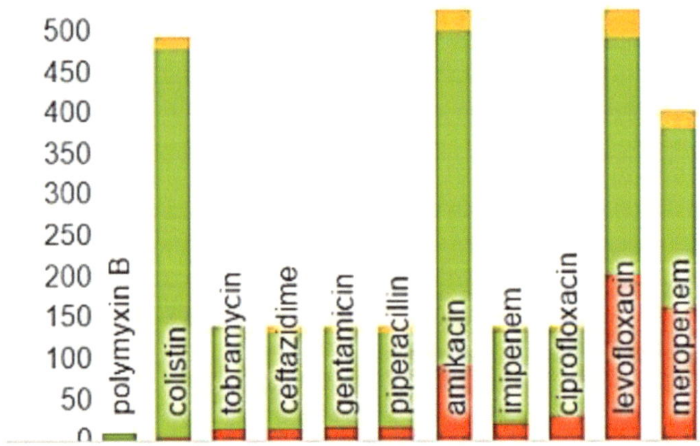

Figure 3.
Strains with P. aeruginosa's genomes showing susceptibility (green) and resistance (red) against some aminoglycosides (amikacin, gentamicin, and tobramycin) and other antipseudomonal antibiotics. Many strains are resistant to a wide range of antibiotics (red with a larger percentage), and most strains are susceptible to doxycycline or colistin (green with larger percentage).

in **Figure 2** antibiotic resistance or susceptibility from different *P. aeruginosa* strains as well as in **Figure 3** with those antibiotics mentioned.

3. Resistance to antibiotics through rRNA methylation

Kgm and Kam families are two different groups of SAM-dependent RNA methyltransferases, which modify nucleotides of 16S rRNAs in the specific drug-binding site to confer self-resistance in aminoglycoside-producing bacteria [31]. The Kgm and Kam families have been distinguished based on their nucleotide targets, G1405 and A1408, respectively. The *kgmB* and *armA* genes (Kgm family kanamycin gentamicin methyltransferase) methylate m^7G1405(N7) position that confers a high level of resistance against gentamicin, kanamycin, and tobramycin. The addition of a methyl group in this position interferes directly with the binding to the antibiotic, inducing a steric hindrance between the modified base and the structure of the antibiotic, causing electrostatic repulsions derived from the positive charge in the modified base [32]. On the other hand, the *kamA* and *npmA* genes (Kam family kanamycin-apramycin methyltransferase) methylate m^1A1408(N1) position conferring a high level of resistance to kanamycin, apramycin, and neomycin [5].

Another interesting non-aminoglycoside resistance related to RNA methylation is the macrolide-lincosamide-streptogramin-B (MLSB) antibiotics, which is strongly associated with the expression of the methyltransferase of ErmC RNA that causes the dimethylation of the N-6 atom of adenine and interacts with the nucleotide 2058 in the 23S rRNA. Such antibiotics bind to overlap sites within the 50S ribosomal subunit tunnel near the peptidyl transferase center, either by inhibiting the catalysis directly at the peptidyl transferase site or by acting as a physical barrier to the extension of the peptide chain inside the tunnel [33]. Many more *erm*-type methyltransferase genes have been identified in a wide range of Gram-positive and Gram-negative bacteria. Among them, the *ermB*, *ermF*, and *ermA* genes are transferred by transposons, and the *ermC* gene transferred by plasmids. The family of Erm methyltransferases that mediate the mono- or dimethylation of A2058 consists of approximately 40 different classes of methylases [34, 35].

4. Resistance or susceptibility to antibiotics through rRNA demethylation

The *rsmG* gene encodes a 16S rRNA mRNA which methylates the N7 of nucleotide G527 within the 530 loop of 16S rRNA; one of the main examples is the loss of native methylations that confers a resistance phenotype to streptomycin. Streptomycin interacts with the rRNA in the adduced region (loop 530), and the loss of methylation correlates with a low level of resistance. Although this resistance is at a low dose of antibiotic, the mutation of *rsmG* apparently has a mutator effect which promotes the appearance of a high number of mutants resistant to high doses of streptomycin [36]. Another interesting aspect is that not only methylation generates resistance; cases have been reported where demethylation also promotes resistance. The first and best characterized example is *ksgA* gene (RsmA protein), which encodes the native methyltransferase KsgA or RsmA, responsible for the N6 dimethylation of A1518 and A1519 in the 3′-terminal fork of the 16S rRNA in the 30S rRNA. It was the first resistance to aminoglycosides (kasugamycin) associated with demethylation in the 16S rRNA [37]. It was found that adenine methylated by MTase is far from the binding site of kasugamycin, so this demethylation should lead to a conformational rearrangement which would be associated with the acquisition of antibiotic resistance [38].

Another research showed that preventing adenine methylation from occurring, resistance to kasugamycin can be induced; the base U793 fills the site usually occupied by the methylated adenines and the adjacent bases, A792 and A794, [39]. The phenomena mentioned above give place to a conformational change, causing the union site of Ksg to be blocked by the U793. Accordingly, it can be assured that this structural change in the helix 24 causes resistance to Ksg [39].

Likewise, it was found that the *tlyA* gene in *Mycobacterium tuberculosis* encodes the MTase 2'-O-ribose TlyA responsible for the C1409 methylations in the 16S rRNA and C1920 in the 23S rRNA. The loss of such methylations confers resistance to capreomycin and viomycin, two antibiotics which bind at the interface of the ribosome subunit and are used to help define their binding site. Another example of the absence of methylations in the 23S rRNA is the lack of methylation in U2584 (*E. coli* numbering), which causes resistance to sparsomycin in 23S rRNA *Halobacterium salinarum* [40].

Recent findings regarding intrinsic resistance refer to the Ψ at position 2504 of the 23S rRNA in *E. coli*, where inactivation of the *rluC* gene confers significant resistance to clindamycin, linezolid, and tiamulin [41]. The *cfr* gene was originally discovered in an isolate of a multiresistant plasmid during a follow-up study of chloramphenicol resistance in *Staphylococcus* spp. isolates. The molecular characterization of the resistance led to the gene encoding a methyltransferase that methylated the nucleotide A2503 in the 23S rRNA. In *E. coli* and *S. aureus*, there is a natural methylation of A2503 mediated by the methyltransferase encoded by the *yfgB* gene (*rlmN*). The lack of natural methylation in A2503 confers a slight increase in susceptibility to tiamulin, hygromycin A, sparsomycin, and linezolid [42].

5. rRNA methyltransferases associated with aminoglycoside resistance in *P. aeruginosa*

Methyltransferases have been intensely studied in *P. aeruginosa*, but this is not the case for RNA methyltransferases, particularly those conferring aminoglycoside resistance. Nowadays, we focus our study in *P. aeruginosa* methyltransferases using molecular biology, genomics, proteomics, chemistry informatics, and bioinformatics [43–45]. RsmG, RsmH, and RsmI are RNA methyltransferases, and these have been broadly studied. Six crystal structures have been reported in PDB for RsmG, from *Thermus thermophilus* with accession numbers 4NXM, 4NXN, 3G88, 3G89 3G8A, and 3G8B [46, 47] and one from *E. coli*, with number 1JSX [48], and another one from *Bacillus subtilis*, with number 1XDZ. RsmH and RsmI crystal structures from *E. coli* are reported in PDB with numbers 3TKA and 5HW4 [49–51]. Checking these three orthologous genes in the PATRIC database, they are being conserved in *P. aeruginosa*'s pan-genome.

RsmG well known as 16S rRNA (guanine^{527}-N^{7})-methyltransferase methylates guanine^{527} at N^{7} in 16S rRNA [36, 52] (**Table 1**) and catalyzes S-adenosyl-L-methionine + guanine^{527} in 16S rRNA → S-adenosyl-L-homocysteine + N^{7}-methylguanine527 in 16S rRNA (see reaction in UniProt, KEGG, or MetaCyc). Researches in *M. tuberculosis* reveal that *rsmG* mutations confer low-level streptomycin resistance; moreover, it has been reported that combining drug resistance mutations of *rsmG* gene remarkably enhances enzyme production in *Paenibacillus agaridevorans* [53]. Likewise, for *P. aeruginosa*, *rsmG* is conserved in both aminoglycoside-resistant and aminoglycoside-susceptible strains.

RsmH also called S-adenosyl-L-methionine (cytosine^{1402}-N^{4})-methyltransferase methylates the N^{4}-of cytosine^{1402} [54] (**Table 1**). This enzyme catalyzes the following chemical reaction: S-adenosyl-L-methionine + cytosine^{1402} in 16S rRNA → S-adenosyl-L-homocysteine + N^{4}-methylcytosine^{1402} in 16S rRNA (see reaction in UniProt, KEGG, or MetaCyc). Experiments performed with gene knockout of *rsmH* and *rsmI* have shown in *E. coli* BW25113 strain that *ΔrsmH* and *ΔrsmI* increase in doubling times by 15 and 12%, respectively; however, *ΔrsmH/ΔrsmI* increases in doubling time by 29% compared with a wild type cultured at 37°C, indicating that gene knockout caused a slight but significant change in phenotype about cellular growth properties in the absence of both *rsmH* and *rsmI* [54]. As well as *E. coli*, *P. aeruginosa* conserves *rsmH* and *rsmI* genes in both aminoglycoside-resistant and susceptible strains; therefore, it is important to study the mutations also in its strains.

RsmI also named S-adenosyl-L-methionine 16S rRNA (cytidine^{1402}–2'-O)-methyltransferase methylates in cytidine^{1402}–2'-O (**Table 1**). RsmI catalyzes the next chemical reaction: S-adenosyl-L-methionine + cytidine^{1402} in 16S rRNA → S-adenosyl-L-homocysteine + 2'-O-methylcytidine^{1402} (see reaction in UniProt, KEGG, or MetaCyc). RsmI and RsmH react on the same nucleotide, but the first methylates in 2'-O, while the second one in -N^{4} [54]. Such as *rsmG* and *rsmH*, the *rsmI* gene is also conserved in pan-genome. Theoretical modeling of the structure in RsmI protein from *P. aeruginosa* was performed in iTISSER suit [55], and compared with 5HW4 from *E. coli* (**Figure 4**), the homology and the active site in *P. aeruginosa* are apparently well maintained.

Other interesting *P. aeruginosa* methyltransferases associated with aminoglycoside resistance are m^{5}C1404, m^{1}A1408, and m^{7}G1405 [6]. Among the last group mentioned, there are some well-studied methyltransferases, such as ArmA, RmtA, RmtB, RmtC, RmtD, RmtF, and RmtG (**Table 2**). This group is characterized for providing resistance to 4,6-disubstituted 2-deoxystreptamine (2-DOS) aminoglycosides [6]. For example, ArmA was found in *Klebsiella pneumoniae* [56]; as for *P. aeruginosa*, among 100 Korean multidrug-resistant isolates, 14 carried this enzyme [57]. The *armA* gene encodes for 16S RNA methyltransferase that methylates guanine (1405)-N^{7}. The same gene in *P. aeruginosa* (**Table 2**) presents variable occurrence as it is part of the accessory genome. A multiple alignment, using the listed P. aeruginosa ArmA proteins (16S rRNA (guanine (1405)-N(7))- methyltransferase)) revealed identical homology for this marker.

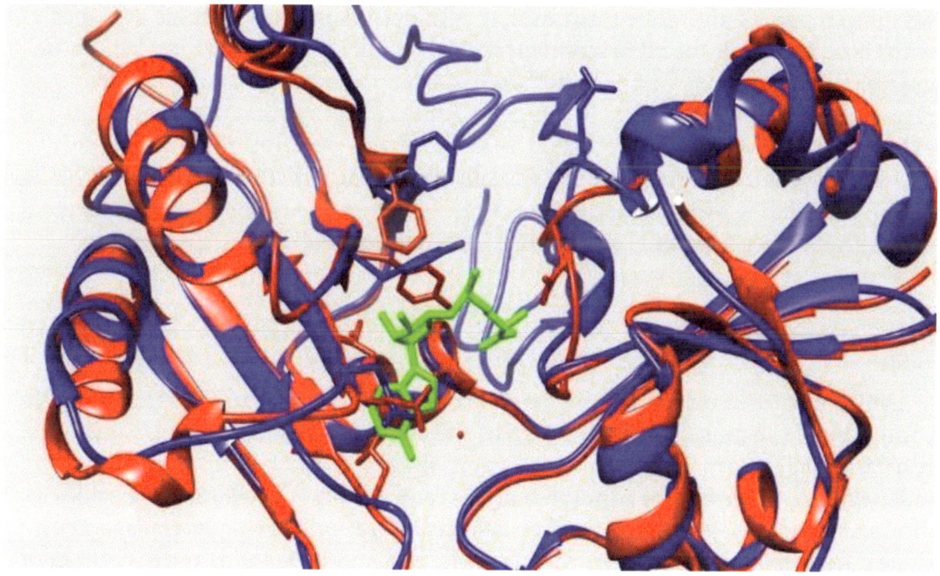

Figure 4.
Structural 3D aligned with chimera 3.1 suit between RsmI proteins from P. aeruginosa (red) and E. coli (blue). Although the alignment displays a structural shifting, the overall topology of the active site is maintained. The protruding residues from each protein depict the active site, harboring SAM (fluorescent green). The structure of E. coli has already been solved by Zhao et al. [49], without obtaining, until now, the crystal structure of RsmI from P. aeruginosa.

Protein	Modification	Strain location	Reference
ArmA	16S rRNA m7G1405	China and Korea	[57, 73]
RmtA	16S rRNA m7G1405	Japan and Korea	[1, 58, 74, 75]
RmtB	16S rRNA m7G1405	China and India	[73, 76]
RmtC	16S rRNA m7G1405	India	[61, 76]
RmtD	16S rRNA m7G1405	Brazil	[77, 78, 79]
RmtF	16S rRNA m7G1405	India	[61, 75]
RmtG	16S rRNA m7G1405	Brazil	[62]

Table 2.
Different 16S rRNA methyltransferases associated with aminoglycoside resistance reported for P. aeruginosa, with classical name, for everyone with the same nucleotide (16S rRNA m7G1405) of isolates from patients who belong to India, Brazil, China, and Korea.

The location of m^7G1405 methyltransferase genes across the prokaryotic genome is variable, as it has been found in several studies mentioned above. The *rmtA* gene in *P. aeruginosa* carries a mobile element Tn5041 [58, 59] identified previously in *Enterobacteriaceae* [56], while the *rmtB* gene identified in *Serratia marcescens* is located in the flanking of Tn3-like region responsible for multiple antimicrobial resistance [59, 60], and both methyltransferases and mobile elements are present in *P. aeruginosa* (**Table 2**). The *rmtC* and *rmtF* genes (**Table 2**) might have been acquired from plasmids as part of mobile genetic elements and finally integrated and stabilized on the chromosome [61]. The *rmtG* gene (**Table 2**) is likely located in the chromosome [62]. The m^1A1408 methyltransferases are present in pan-aminoglycoside-resistant strains, which were identified by Wachino et al. [63] and provide resistance by these classes of methyltransferases to both 4,5-disubstituted 2-DOS and 4,6-disubstituted 2-DOS aminoglycosides as well as NpmA case. These two classes of methyltransferases are very important for antibiotic resistance, thanks to the similarity of these enzymes with those homologs found in aminoglycoside-producing actinomycetes [6]. Those new genes and proteins will be better studied in expression and structure, to be related to epidemiological data. Looking at **Table 2**, it seems that the expression of methyltransferase might be related to geographical prescription. However, this hypothesis does not seem well founded: microbiological, molecular, and epidemiological understanding of RNA methyltransferases in *P. aeruginosa* will allow the rational use of aminoglycosides and maybe will be not replaced for new antibiotics.

6. Final considerations

Antibiotic resistance is a serious concern for public health and environment. To comprehend the molecular interaction of the methyltransferase in aminoglycoside resistance will be a more efficient way to rationalize its use and consumption. It will be better to clarify the panorama of the rational use of the aminoglycosides to diminish the rapid development of resistance before considering its replacement, since *P. aeruginosa* is still susceptible to them, and, moreover, currently it is known why other *Gammaproteobacteria* are resistant to them. Why are methylation and demethylation a feedback of the antibiotic environmental pressure in bacteria? Bhujbalrao and Anand [64] suggest us some insights using KsgA, exploring the factors which govern the resistance to antibiotics. They observed within loop1 and loop12 of rRNA switched chimera efficiently methylated mini-RNA substrates in vitro, showing that these structural elements suffice for local orientation of the rRNA. In addition, in vivo they notice that the head domain plays a more critical role in leading the enzyme to the select ribosomal region and serves as a sensor of the global environment.

As Kim et al. [65] discuss in letter to editor (Dr. Hur), investigating with P. aeruginosa and aminoglycoside resistance proposes that "less aminoglycoside consumption correlates with less resistance levels"; therefore, we consider that is a requisite for an antibiotic cycling strategy at the global level; also they discussed the rates of amikacin or gentamicin-resistant declining trends, according to the data from KONSAR Korean program in 2011 either for *P. aeruginosa*, *K. pneumoniae*, or *Acinetobacter* spp. [64]. With the knowledge about aminoglycoside resistance molecular mechanisms comparing to the rational prescription cited in Korea, for example, we hypothesize that low methylation rate in the nucleotide substrate of RsmH or RsmI is close to the anchor point of gentamicin in 16S RNA, indicating a possible association with gentamicin or aminoglycosides resistance [5].

RsmG, RsmH, and RsmI methyltransferases belong to the core genome (constitutive genome), while ArmA is part of the accessory genome with identical protein sequences among close species in *Proteobacteria*. Nowadays, the enzymatic activity has been well described; however the antibiotic resistance remains unsolved, perhaps as a consequence of broad usage of aminoglycoside in hospital environment, allowing the development of resistant bacteria. In the future, probably the treatment of *P. aeruginosa* will take into account the genetic trait of each isolate, strain, or species with the set of resistance genes, and surely methyltransferases will be included routinely in clinical care and high throughput or genomic medicine therapies.

Acknowledgements

We thank the University of Antioquia for their financial support to PhD student Jaison H. Cuartas, project CODI 2017-15753.

Author details

Pablo Valderrama-Carmona, Jaison H. Cuartas, Diana Carolina Castaño and Mauricio Corredor*
Gebiomic group, Natural and Exact Sciences Faculty, University of Antioquia, Medellín, Colombia

*Address all correspondence to: mauricio.corredor@udea.edu.co

IntechOpen

© 2019 The Author(s). Licensee IntechOpen. This chapter is distributed under the terms of the Creative Commons Attribution License (http://creativecommons.org/licenses/by/3.0), which permits unrestricted use, distribution, and reproduction in any medium, provided the original work is properly cited.

References

[1] Yokoyama K, Doi Y, Yamane K, Kurokawa H, Shibata N, Shibayama K, et al. Acquisition of 16S rRNA methylase gene in *Pseudomonas aeruginosa*. Lancet. 2003;**362**(9399):1888-1893

[2] Swinehart WE, Jackman JE. Diversity in mechanism and function of tRNA methyltransferases. RNA Biology. 2015;**12**(4):398-411

[3] Schubert HL, Blumenthal RM, Cheng X. Many paths to methyltransfer: A chronicle of convergence. Trends in Biochemical Sciences. 2003;**28**(6):329-335

[4] The RNA Institute, College of Arts and Sciences SU of NY at A. The RNA Modification Database [Internet]. 2018. Available from: https://mods.rna.albany.edu/mods/modifications/view/

[5] Beauclerk AAD, Cundliffe E. Sites of action of two ribosomal RNA methylases responsible for resistance to aminoglycosides. Journal of Molecular Biology. 1987;**193**(4):661-671

[6] Gutierrez B, Douthwaite S, Gonzalez-Zorn B. Indigenous and acquired modifications in the aminoglycoside binding sites of *Pseudomonas aeruginosa* rRNAs. RNA Biology. 2013;**10**(8):1324-1332

[7] Magnet S, Blanchard JS. Molecular insights into aminoglycoside action and resistance. Chemical Reviews. 2005;**105**(2):477-497

[8] Rana AK, Ankri S. Reviving the RNA world: An insight into the appearance of RNA methyltransferases. Frontiers in Genetics. 2016;**7**(99):1-9

[9] Poehlsgaard J, Douthwaite S. The bacterial ribosome as a target for antibiotics. Nature Reviews. Microbiology. 2005;**3**(11):870-881

[10] Conn GL, Savic M, Macmaster R. Antibiotic resistance in bacteria through modification of nucleosides in 16S ribosomal RNA. In: Grosjean H, editor. DNA and RNA Modification Enzymes: Structure, Mechanism, Function and Evolution. Texas, USA: Landes Bioscience; 2009. p. 653

[11] MacMaster R, Zelinskaya N, Savic M, Rankin CR, Conn GL. Structural insights into the function of aminoglycoside-resistance A1408 16S rRNA methyltransferases from antibiotic-producing and human pathogenic bacteria. Nucleic Acids Research. 2010;**38**(21):7791-7799

[12] Castanheira M, Mills JC, Farrell DJ, Jones RN. Mutation-driven β-lactam resistance among contemporary ceftazidime non-susceptible *Pseudomonas aeruginosa* isolates from USA hospitals. Antimicrobial Agents and Chemotherapy. 2014;**58**(11):6844-6850

[13] Livermore DM, Mushtaq S, Ge Y, Warner M. Activity of cephalosporin CXA-101 (FR264205) against *Pseudomonas aeruginosa* and *Burkholderia cepacia* group strains and isolates. International Journal of Antimicrobial Agents. 2009;**34**(5):402-406

[14] Livermore DM. β-Lactamases in laboratory and clinical resistance. Clinical Microbiology Reviews. 1995;**8**(4):557-584

[15] Jalal S, Wretlind B. Mechanisms of quinolone resistance in clinical strains of *Pseudomonas aeruginosa*. Microbial Drug Resistance. 1998;**4**(4):257-261. Available from: http://www.liebertpub.com/doi/10.1089/mdr.1998.4.257

[16] Poole K. Efflux-mediated multiresistance in Gram-negative bacteria. European Society of Clinical

Infectious Diseases. 2004;**10**(1):12-26. DOI: 10.1111/j.1469-0691.2004.00763.x

[17] Poole K. Efflux pumps as antimicrobial resistance mechanisms. Annals of Medicine. 2007;**39**(3):162-176

[18] Strateva T, Yordanov D. *Pseudomonas aeruginosa*—A phenomenon of bacterial resistance. Journal of Medical Microbiology. 2009;**58**(9):1133-1148

[19] Li X-Z, Poole K, Nikaido H. Contributions of MexAB-OprM and an EmrE homolog to intrinsic resistance of *Pseudomonas aeruginosa* to aminoglycosides and dyes. Antimicrobial Agents and Chemotherapy. 2003;**47**(1):27-33

[20] Shu JC, Chia JH, Siu LK, Kuo AJ, Huang SH, Su LH, et al. Interplay between mutational and horizontally acquired resistance mechanisms and its association with carbapenem resistance amongst extensively drug-resistant *Pseudomonas aeruginosa* (XDR-PA). International Journal of Antimicrobial Agents. 2012;**39**(3):217-222. DOI: 10.1016/j.ijantimicag.2011.09.023

[21] Li XZ, Zhang L, Poole K. Interplay between the MexA-MexB-OprM multidrug efflux system and the outer membrane barrier in the multiple antibiotic resistance of *Pseudomonas aeruginosa*. The Journal of Antimicrobial Chemotherapy. 2000;**45**(4):433-436

[22] Nikaido H. Multidrug resistance in bacteria. Annual Review of Biochemistry. 2009;**78**:119-146

[23] Bateman A, Martin MJ, O'Donovan C, Magrane M, Apweiler R, Alpi E, et al. UniProt: A hub for protein information. Nucleic Acids Research. 2015;**43**(D1):D204-D212

[24] Mena A, Smith EE, Burns JL, Speert DP, Moskowitz SM, Pérez JL, et al. Genetic adaptation of *Pseudomonas aeruginosa* to the airways of cystic fibrosis patients is catalyzed by hypermutation. Journal of Bacteriology. 2008;**190**(24):7910-7917

[25] Khosravi AD, Motahar M, Montazeri EA. The frequency of class1 and 2 integrons in *Pseudomonas aeruginosa* strains isolated from burn patients in a burn center of Ahvaz, Iran. PLoS One. 2017;**12**(8):e0183061

[26] Fluit AC, Schmitz FJ. Resistance integrons and super-integrons. Clinical Microbiology and Infection. 2004;**10**(4):272-288. DOI: 10.1111/j.1198-743X.2004.00858.x

[27] Gillings MR. Integrons: Past, present, and future. Microbiology and Molecular Biology Reviews. 2014;**78**(2):257-277

[28] Bissonnette L, Champetier S, Buisson JP, Roy PH. Characterization of the nonenzymatic chloramphenicol resistance (cmlA) gene of the In4 integron of Tn1696: Similarity of the product to transmembrane transport proteins. Journal of Bacteriology. 1991;**173**(14):4493-4502

[29] Mirsalehian A, Feizabadi M, Nakhjavani FA, Jabalameli F, Goli H, Kalantari N. Detection of VEB-1, OXA-10 and PER-1 genotypes in extended-spectrum β-lactamase-producing *Pseudomonas aeruginosa* strains isolated from burn patients. Burns. 2010;**36**(1):70-74

[30] Gillespie JJ, Wattam AR, Cammer SA, Gabbard JL, Shukla MP, Dalay O, et al. PATRIC: The comprehensive bacterial bioinformatics resource with a focus on human pathogenic species. Infection and Immunity. 2011;**79**(11):4286-4298

[31] Savic M, Lovrić J, Tomic TI, Vasiljevic B, Conn GL. Determination of the target nucleosides for members of two families of 16s rRNA methyltransferases that

confer resistance to partially overlapping groups of aminoglycoside antibiotics. Nucleic Acids Research. 2009;**37**(16):5420-5431

[32] Tomic TI, Moric I, Conn GL, Vasiljevic B. Aminoglycoside resistance genes sgm and kgmB protect bacterial but not yeast small ribosomal subunits in vitro despite high conservation of the rRNA A-site. Research in Microbiology. 2008;**159**(9-10):658-662. DOI: 10.1016/j.resmic.2008.09.006

[33] Skinner R, Cundliffe E, Schmidt FJ. Site of action of a ribosomal RNA methylase responsible for resistance to erythromycin and other antibiotics. The Journal of Biological Chemistry. 1983;**258**(20):12702-12706

[34] Roberts MC, Sutcliffe J, Courvalin P, Jensen LB, Rood J, Seppala H. Nomenclature for macrolide and macrolide-lincosamide-streptogramin B resistance determinants. Antimicrobial Agents and Chemotherapy. 1999;**43**(12):2823-2830

[35] Roberts MC. Update on macrolide-lincosamide-streptogramin, ketolide, and oxazolidinone resistance genes. FEMS Microbiology Letters. 2008;**282**(2):147-159

[36] Okamoto S, Tamaru A, Nakajima C, Nishimura K, Tanaka Y, Tokuyama S, et al. Loss of a conserved 7-methylguanosine modification in 16S rRNA confers low-level streptomycin resistance in bacteria. Molecular Microbiology. 2007;**63**(4):1096-1106

[37] Helser TL, Davies JE, Dahlberg JE. Mechanism of Kasugamycin resistance in *Escherichia coli*. Nature: New Biology. 1972;**235**:(6):226:229

[38] Schuwirth BS, Borovinskaya MA, Hau CW, Zhang W, Vila-Sanjurjo A, Holton JM, et al. Structures of the bacterial ribosome at 3.5 Å resolution. Science (80-). 2005;**310**(5749):827-834

[39] Demirci H, Murphy F IV, Belardinelli R, Kelley AC, Ramakrishnan V, Gregory ST, et al. Modification of 16S ribosomal RNA by the KsgA methyltransferase restructures the 30S subunit to optimize ribosome function. RNA. 2010;**16**(12):2319-2324

[40] Lázaro E, Rodriguez-Fonseca C, Porse B, Ureña D, Garrett RA, Ballesta JPG. A sparsomycin-resistant mutant of *Halobacterium salinarium* lacks a modification at nucleotide U2603 in the peptidyl transferase centre of 23 S rRNA. Journal of Molecular Biology. 1996;**261**:231-238

[41] Toh SM, Mankin AS. An indigenous posttranscriptional modification in the ribosomal peptidyl transferase center confers resistance to an array of protein synthesis inhibitors. Journal of Molecular Biology. 2008;**380**(4):593-597

[42] Toh SM, Xiong L, Bae T, Mankin AS. The methyltransferase YfgB/RlmN is responsible for modification of adenosine 2503 in 23S rRNA. RNA. 2008;**14**(1):98-106

[43] Mosquera-Rendón J, Cárdenas-Brito S, Pineda JD, Corredor M, Benítez-Páez A. Evolutionary and sequence-based relationships in bacterial AdoMet-dependent non-coding RNA methyltransferases. BMC Research Notes. 2014;**7**(1):440

[44] Mosquera-Rendón J, Rada-Bravo AM, Cárdenas-Brito S, Corredor M, Restrepo-Pineda E, Benítez-Páez A. Pangenome-wide and molecular evolution analyses of the *Pseudomonas aeruginosa* species. BMC Genomics. 2016;**17**(1):45. DOI: 10.1186/s12864-016-2364-4

[45] Castaño DC, Mosquera-Redón J, Corredor M. Structural and functional genomics analysis of methyltransferase genes and networks associate to understand antibiotic resistance inside the pangenome of *Pseudomonas*

aeruginosa. In: VII Latin American Congress on Biomedical Engineering CLAIB 2016; Bucaramanga, Santander, Colombia: Springer, Singapore. 2016. pp. 702-705

[46] Demirci H, Gregory ST, Dahlberg AE, Jogl G. Crystal structure of the *Thermus thermophilus* 16 S rRNA methyltransferase RsmC in complex with cofactor and substrate guanosine. The Journal of Biological Chemistry. 2008;**283**(39):26548-26556

[47] Gregory ST, Demirci H, Belardinelli R, Monshupanee T, Gualerzi C, Dahlberg AE, et al. Structural and functional studies of the *Thermus thermophilus* 16S rRNA methyltransferase RsmG. RNA. 2009;**15**(9):1693-1704

[48] Romanowski MJ, Bonanno JB, Burley SK. Crystal structure of the *Escherichia coli* glucose-inhibited division protein B (GidB) reveals a methyltransferase fold. Proteins: Structure, Function, and Genetics. 2002;**47**(4):563-567

[49] Zhao M, Wang L, Zhang H, Dong Y, Gong Y, Zhang L, et al. Purification, crystallization and preliminary crystallographic analysis of the 16S rRNA methyltransferase RsmI from *Escherichia coli*. Acta Crystallographica Section F: Structural Biology Communications. 2014;**70**(9):1256-1259

[50] Wei Y, Zhang H, Gao ZQ, Wang WJ, Shtykova EV, Xu JH, et al. Crystal and solution structures of methyltransferase RsmH provide basis for methylation of C1402 in 16S rRNA. Journal of Structural Biology. 2012;**179**(1):29-40. DOI: 10.1016/j.jsb.2012.04.011

[51] Zhao M, Zhang H, Liu G, Wang L, Wang J, Gao Z, et al. Structural insights into the methylation of C1402 in 16S rRNA by methyltransferase RsmI. PLoS One. 2016;**11**(10):1-15

[52] Schomburg D, Schomburg I, Chang A. In: Schomburg D, Schomburg I, Chang A, editors. Class 2-3.2 Transferases, Hydrolases: EC 2-3.2. Second. Braunschweig, Germany: Springer Science & Business Media, 2013. 698 p

[53] Funane K, Tanaka Y, Hosaka T, Murakami K, Miyazaki T, Shiwa Y, et al. Combined drug resistance mutations substantially enhance enzyme production in *Paenibacillus agaridevorans*. Journal of Bacteriology. 2018;**200**(17):e00118-e00188

[54] Kimura S, Suzuki T. Fine-tuning of the ribosomal decoding center by conserved methyl-modifications in the *Escherichia coli* 16S rRNA. Nucleic Acids Research. 2010;**38**(4):1341-1352

[55] Yang J, Yan R, Roy A, Xu D, Poisson J, Zhang Y. The I-TASSER suite: Protein structure and function prediction. Nature Methods. 2015;**12**(1):7-8. Available from: http://www.nature.com/doifinder/10.1038/nmeth.3213

[56] Galimand M, Courvalin P, Lambert T. Plasmid-mediated high-level resistance to aminoglycosides in Enterobacteriaceae due to 16S rRNA methylation. Antimicrobial Agents and Chemotherapy. 2003;**47**(8):2565-2571

[57] Gurung M, Moon DC, Tamang MD, Kim J, Lee YC, Seol SY, et al. Emergence of 16S rRNA methylase gene armA and cocarriage of blaIMP-1in *Pseudomonas aeruginosa* isolates from South Korea. Diagnostic Microbiology and Infectious Disease. 2010;**68**(4):468-470. DOI: 10.1016/j.diagmicrobio.2010.07.021

[58] Yamane K, Doi Y, Yokoyama K, Yagi T, Kurokawa H, Shibata N, et al. Genetic environments of the rmtA gene in *Pseudomonas aeruginosa* clinical isolates. Antimicrobial Agents and Chemotherapy. 2004;**48**(6):2069-2074

[59] Yamane K, Wachino JI, Doi Y, Kurokawa H, Arakawa Y. Global spread of multiple aminoglycoside resistance genes. Emerging Infectious Diseases. 2005;**11**(6):951-953

[60] Doi Y, Yokoyama K, Yamane K, Wachino J, Shibata N, Yagi T, et al. Plasmid-mediated 16S rRNA methylase in *Serratia marcescens* conferring high-level resistance to aminoglycosides. Antimicrobial Agents and Chemotherapy. 2004;**48**(2):491-496

[61] Rahman M, Prasad KN, Pathak A, Pati BK, Singh A, Ovejero CM, et al. RmtC and RmtF 16S rRNA methyltransferase in NDM-1–producing *Pseudomonas aeruginosa*. Emerging Infectious Diseases. 2015;**21**(11):2059-2062

[62] Francisco GR, Nora STR, Bueno MFC, Da Silva Filho LVRF, De Oliveira Garcia D. Identification of aminoglycoside-resistant *Pseudomonas aeruginosa* producing RmtG 16S rRNA methyltransferase in a cystic fibrosis patient. Antimicrobial Agents and Chemotherapy. 2015;**59**(5):2967-2968

[63] Wachino JI, Shibayama K, Kurokawa H, Kimura K, Yamane K, Suzuki S, et al. Novel plasmid-mediated 16S rRNA m1A1408 methyltransferase, NpmA, found in a clinically isolated *Escherichia coli* strain resistant to structurally diverse aminoglycosides. Antimicrobial Agents and Chemotherapy. 2007;**51**(12):4401-4409

[64] Bhujbalrao R, Anand R. Deciphering determinants in ribosomal methyltransferases that confer antimicrobial resistance. Journal of the American Chemical Society. 2019. Available from: http://pubs.acs.org/doi/10.1021/jacs.8b10277

[65] Kim YA, Park YS, Youk T, Lee H, Lee K. Correlation of aminoglycoside consumption and amikacin- or gentamicin-resistant *Pseudomonas aeruginosa* in long-term nationwide analysis: Is antibiotic cycling an effective policy for reducing antimicrobial resistance? Annals of Laboratory Medicine. 2018;**38**(2):176-178

[66] Starr JL, Fefferman R. The occurrence of methylated bases in ribosomal ribonucleic acid of *Escherichia coli* K12 W-6. Journal of Biological Chemistry. 1964;**239**(10):3457-3461. Available from: http://www.ncbi.nlm.nih.gov/pubmed/14245403

[67] Hayashi Y, Osawa S, Miura K. The methyl groups in ribosomal RNA from *Escherichia coli*. Biochimica et Biophysica Acta. 1966;**129**:519-531

[68] Dubin DT, Gunalp A. Minor nucleotide composition of ribosomal precursor, and ribosomal, ribonucleic acid in *Escherichia coli*. Biochimica et Biophysica Acta. 1967;**134**:106-123

[69] Nichols JL, Lane BG. N4-methyl-2′-O-methyl cytidine and other methyl-substituted nucleoside constituents of *Escherichia coli* ribosomal and soluble RNA. Biochimica et Biophysica Acta. 1966;**119**:649-651

[70] Isaksson LA, Phillips JH. Studies on microbial RNA. V. A comparison of the in vivo methylated components of ribosomal RNA from *Escherichia coli* and *Saccharomyces cerevisiae*. Biochimica et Biophysica Acta. 1968;**155**:63-71

[71] Fellner P. Nucleotide sequences from specific areas of the 16S and 23S ribosomal RNAs of *E. coli*. European Journal of Biochemistry. 1969;**11**(1):12-27

[72] Johnson JD, Horowitz J. Characterization of ribosomes and RNAs from *Mycoplasma hominis*. Biochimica et Biophysica Acta. 1971;**247**:262-279

[73] Zhou Y, Yu H, Guo Q, Xu X, Ye X, Wu S, et al. Distribution of 16S rRNA

methylases among different species of gram-negative bacilli with high-level resistance to aminoglycosides. European Journal of Clinical Microbiology & Infectious Diseases. 2010,**29**(11).1349-1353

[74] Yamane K, Wachino J, Suzuki S, Shibata N, Kato H, Shibayama K, et al. 16S rRNA methylase-producing, Gram-negative pathogens, Japan. Emerging Infectious Diseases. 2007;**13**(4):642-646

[75] Jin JS, Kwon KT, Moon DC, Lee JC. Emergence of 16S rRNA methylase rmtA in colistin-only-sensitive *Pseudomonas aeruginosa* in South Korea. International Journal of Antimicrobial Agents. 2009;**33**(5):490-491

[76] Mohanam L, Menon T. Emergence of rmtC and rmtF 16S rRNA methyltransferase in clinical isolates of *Pseudomonas aeruginosa*. Indian Journal of Medical Microbiology. 2017;**35**(2):282-285

[77] Doi Y, De Oliveira Garcia D, Adams J, Paterson DL. Coproduction of novel 16S rRNA methylase RmtD and metallo-β-lactamase SPM-1 in a panresistant *Pseudomonas aeruginosa* isolate from Brazil. Antimicrobial Agents and Chemotherapy. 2007;**51**(3): 852-856

[78] Castanheira M, Fritsche TR, Sader HS, Jones RN. RmtD 16S RNA Methylase in epidemiologically unrelated SPM-1-producing *Pseudomonas aeruginosa* isolates from Brazil. Antimicrobial Agents and Chemotherapy. 2008;**52**(4):1587-1588

[79] Lincopan N, Neves P, Mamizuka EM, Levy CE. Balanoposthitis caused by *Pseudomonas aeruginosa* co-producing metallo-β-lactamase and 16S rRNA methylase in children with hematological malignancies. International Journal of Infectious Diseases. 2010;**14**(4):e344-e347. DOI: 10.1016/j.ijid.2009.04.016

Chapter 5

Bulgecins as β-Lactam Enhancers Against Multidrug Resistant (MDR) *Pseudomonas aeruginosa*

Marion J. Skalweit

Abstract

Antibiotic resistance in non-lactose fermenting pathogens such as *Pseudomonas aeruginosa* (*P. aeruginosa*) is increasing, making these clinical pathogens more difficult to treat. Multiple resistance mechanisms exist within *P. aeruginosa* that affect all classes of antibiotics used in the clinic. New strategies and treatment targets within these MDR pathogens must be exploited. One heretofore untapped target is the family of cell wall enzymes known as lytic transglycosylases (Lts). Lts work in concert with penicillin binding proteins (PBPs) and other cell wall proteins such as amidases and peptidoglycan hydrolases to affect normal cell division, and during stress and programmed cell death. Lts are inhibited by natural products called bulgecins, produced by non-pathogenic *Paraburkholderia* and *Burkholderia* spp. New research describing the ability of Lt inhibition to restore susceptibility to β-lactams in MDR *P. aeruginosa*, as well as the structural biologic basis for the activity of bulgecins will be reviewed. Other targets and applications of bulgecins will also be discussed.

Keywords: antibiotic resistance, *Pseudomonas aeruginosa*, metallo-β-lactamase, penicillin binding protein, lytic transglycosylase, bulgecin A

1. Introduction

This chapter will review our current state of the art knowledge about bulgecins, natural inhibitors of lytic transgylcosylase cell wall enzymes, and their activity as β-lactam enhancers to inhibit growth of *P. aeruginosa*. Current known resistance mechanisms targeting β-lactams in *P. aeruginosa* be reviewed, followed by an introduction to the lytic transglycosylases of *P. aeruginosa*. The use of bulgecins as adjunctive agents with β-lactams will be described as well as the synthesis of Bulgecin A, the most active of these compounds.

2. β-Lactam resistance mechanisms in *Pseudomonas aeruginosa*

P. aeruginosa has intrinsically higher minimum inhibitory concentrations (MICs) against anti-pseudomonal β-lactams when compared to Enterobacteriaceae, even in the absence of specific resistance determinants. For example, typical MICs for the anti-pseudomonal cephalosporin, ceftazidime, for *Escherichia coli* are 0.06–0.125 mg/L whereas for *P. aeruginosa* isolates, MICs are in the range of 1–2 mg/L. In the main, when *P. aeruginosa* is resistant to β-lactams, specific mechanisms are at play. These include downregulation of outer membrane porins, expression of intrinsic efflux mechanisms, acquisition of β-lactamase enzymes including carbapenamases such as IMP ("imipenemase"), VIM ("Verona imipenemase") and KPC ("Klebsiella pneumoniae carbapenemase"), hyperproduction of chromosomal β-lactamases ("AmpCs or Pseudomonas-Derived Cephalosporinases, "PDCs") and reduced penicillin binding protein affinity for β-lactams that are not considered "anti-pseudomonal" β-lactams. **Table 1** summarizes the mechanism and resistance determinants responsible (adapted from [1]).

Currently, in the clinical microbiology laboratory, susceptibilities are reported to particular antibiotics depending on the specific sample submitted, e.g., urine, blood, sterile body fluids (pleural, joint, cerebrospinal fluid). At least initially, genotypic testing to determine the presence of specific antibiotic resistance determinants is not performed, and it is left to the clinical infectious diseases expert to reason out the most likely resistance mechanisms based on susceptibility patterns, and to select the most appropriate antibiotic(s) for treatment.

2.1 Outer membrane porin loss (OprD)

The structure, function and regulation of *P. aeruginosa* porins is complex and has been recently reviewed [2]. Porins are involved in structural and signaling tasks in *P. aeruginosa*, as well as passage of nutrients. The Opr D family of porins is the largest and is subdivided into two groups OccD and OccK. These porins are each regulated through their own sigma factors. Porin loss can be related to formation of OprD containing outer membrane vesicles that are also found in high concentrations in biofilms. Resistance to carbapenems in biofilms may be related to this. Mutations in *P. aeruginosa* that cause oprD (occD1) to not be expressed are linked to imipenem resistance. Other occD1 mutations that do not effect transcription also lead to carbapenem resistance [3, 4]. OprD mutations or loss is often associated with overexpression of efflux pumps (see below) leading to high level resistance to carbapenems, other β-lactams and other classes of antibiotics such as aminoglycosides and fluoroquinolones [5].

In the clinical setting, OprD porin loss is often associated with a resistance phenotype in which one observes resistance to carbapenems including imipenem, but *in vitro* susceptibility to anti-pseudomonal pencillins and cephalosporins. In

	Resistance determinant	Antibiotics affected
Pseudomonas aeruginosa	OprD loss	Carbapenems, some cephalosporins, penicillins
	Efflux pumps (e.g., MexA-B/OprM)	Meropenem, some cephalosporins, penicillins
	Chromosomal Amp C of *P. aeruginosa*	Anti *P. aeruginosa* penicillins, anti *P. aeruginosa* cephalosporins except ceftolozane
	Acquired ESBLs (TEM, SHV,OXA, GES, VEB, CTX-M, PER)	anti *P. aeruginosa* cephalosporins except ceftolozane, cefepime
	Acquired carbapenemases (KPC, OXA, metallo-β-lactamases like NDM, VIM, IMP, SPM types)	anti *P. aeruginosa* pencillins, cephalosporins and carbapenems

OprD, outer membrane porin D; Mex-multidrug efflux; TEM, class A β-lactamase of E. coli, named for patient in which it was discovered; SHV, sulfhydryl variant of TEM; OXA, oxacillinase; GES, German extended spectrum β-lactamase; VEB, Verona extended spectrum; CTX-M, cefotaximase-München; PER, plasmidic extended spectrum β-lactamase; NDM, New Delhi metallo-β-lactamase; SPM, Sao Paolo metallo-β-lactamase.

Table 1.
β-Lactam resistance determinants in P. aeruginosa.

combination with other resistance mechanisms, such as over-expression of the chromosomal PDC enzymes, or presence of other acquired cephalosporinases such as TEM, SHV and OXA β-lactamases, higher level carbapenem resistance is observed as well as resistance to other classes of β-lactam antibiotics.

2.2 Efflux pumps in *P. aeruginosa*

As with porins, *P. aeruginosa* possesses a large variety of efflux pumps that perform different roles in the bacteria, but mainly function to extrude harmful substances from the cell. These pumps have been reviewed in [6]. Pumps of importance in carbapenem and other antibiotic efflux are in the resistance nodulation or RND type family and include the MexAB-OprM, MexCD-OprJ, MexEF-OprN and MexXY-OprM multidrug efflux pump systems [7]. Increased expression of these pumps leads to high level carbapenem resistance, often in association with OprD loss or modification. Notably imipenem is not a substrate of the multidrug efflux pumps of *P. aeruginosa* while meropenem is [7]. In the clinical setting, if one notes resistance to meropenem and other β-lactams, except for imipenem, then an efflux mechanism is at play. If resistance is noted to both meropenem and imipenem, but not to other β-lactams, OprD loss or modification is responsible. With resistance to carbapenems as well as other β-lactams, multiple resistance mechanisms can be at play including efflux, intrinsic and acquired β-lactamases and decreased permeability (porin loss).

2.3 Hyperproduction of PDC β-lactamases

As in other organisms, of which *Enterobacter cloacae* is the most well-known example, *P. aeruginosa* possesses a chromosomal AmpC β-lactamase also called PDC. Chromosomal β-lactamases likely play a role in cell wall maintenance, as well as degradation of β-lactam antibiotics. As characterized in *E. cloacae* [8], the AmpC cephalosporinases are under the regulation of *amp*R, a LysR type regulatory system [9]. Under normal circumstances, there is low level constitutive expression of the AmpC protein. Upon exposure to β-lactam antibiotics, muramyl pentapeptides are released that displace a repressor protein encoded by *amp*R from the promoter of AmpC. This leads to increased expression of AmpC cephalosporinase. The increased expression of AmpC can occur with exposure to cephamycins like cefoxitin for example. Increased expression of AmpC in *E. cloacae* occurs via a pathway involving NagZ, a N-acetyl-β-D-glucosamindase, or independent of NagZ [8]. The muramyl pentapeptides are also degraded by a cytosolic amidase, Amp D. This leads to re-association of the repressor to the promoter and resumption of normal levels of Amp C expression. There are also insertion sequence mutations in AmpR that can lead to increased expression of AmpC, as well as mutations in AmpD amidases that reduce degradation of muramyl pentapeptides. The regulation of Amp Cs differs somewhat in *P. aeruginosa*, involving 2 pathways that include the lytic transglycosylases Slt, SltB1, MltB and MltF, and PBP 4 in the generation of muramyl peptides [10]. Mutations in PBP4 are associated with higher levels of AmpC expression. Finally there are specific AmpC mutations that can lead to a carbapenemase phenotype in these enzymes, although the significance of this in terms of clinically relevant carbapenem resistance is unclear [11].

2.4 Acquired β-lactamases in *P. aeruginosa*

β-lactamases from all four Ambler classes have been described in *P. aeruginosa*, including Class A extended spectrum β-lactamases (ESBLs) of the TEM, SHV,

CTX-M, GES, PER and VEB types; Class A carbapenemases such as KPC variants; metallo-β-lactamases such as the VIM, IMP, NDM and SPM B1 di-Zn^{2+} enzymes: and OXA carbapenemases [9, 12]. Weak imipenemases in the so-called Class C AmpCs have already been discussed above. In combination with OprD loss and/or upregulation of MEX efflux pumps, high level carbapenem resistance can be seen in *P. aeruginosa* due to acquired β-lactamases. Traditional class A β-lactamase inhibitors such as clavulanic acid, sulbactam and tazobactam have *in vitro* activity versus the Class A ESBLs but not against other β-lactamases, e.g., the anti-pseudomonal combination ceftolozane-tazobactam is not effective against KPC, metallo-b-lactamases, or OXA enzymes [1]. New β-lactam-β-lactamase inhibitor combinations such as ceftazidime-avibactam and meropenem-vaborbactam will be active against isolates with KPC enzymes, and Class C β-lactamases, as long as they lack other resistance mechanisms that increase the β-lactam MIC beyond what is caused by the β-lactamase enzyme [13, 14].

2.5 Penicillin binding proteins (PBPs) of *P. aeruginosa*

PBPs of *P. aeruginosa* have high affinities for so called anti-pseudomonal β-lactams namely piperacillin, ticarcillin, ceftazidime, cefepime, ceftolozane, meropenem, imipenem, doripenem and aztreonam [1]. PBP 3 is the most important target of inhibition as it is essential for growth of the bacteria [15]. PBP3 is the primary target for ceftazidime whereas PBP2 is the target of carbapenems. Mutations in PBPs have not been described in *P. aeruginosa* leading to β-lactam resistance. The interactions of specific PBPs with specific lytic transglycosylases in the maintenance of cell wall will be further discussed below.

2.6 Current therapeutic strategies to treat infections with resistant *P. aeruginosa*

Given that 15–33% of *P. aeruginosa* isolates are multidrug resistant (have at least one resistance mechanism) [16, 17] and that resistance is associated with up to fivefold greater mortality [18, 19], choosing the right antibiotic combinations have a tremendous impact on patient outcomes. Advances in the rapid diagnosis of *P. aeruginosa*, and use of both rapid phenotypic tests such as CARBA NP [20] or rapid molecular diagnostics to identify specific ESBL and carbapenemase enzymes, have enhanced the clinician's ability to get patients on the right therapy sooner. Identification of patient risk factors, including prior antibiotic exposure, and knowledge of local trends in resistance patterns are useful in selection of empiric antibiotics, until antimicrobial susceptibilities and genotypic results are available for guidance. Carbapenems (meropenem or imipenem) and anti-pseudomonal cephalosporins in combination with colistin, an aminoglycoside or fosfomycin, versus ceftolozane/tazobactam or meropenem/vaborbactam or ceftazidime/avibactam are all good empiric choices for critically ill patients [16], provided multidrug resistance is not present. However, clearly more therapeutic options are needed for infections with extensively drug resistant and pan-resistant *P. aeruginosa*. Lytic transglycosylases represent a new target for bacterial inhibition.

3. Lytic transglycosylases of *P. aeruginosa*

Recently, lytic transglycosylases of *P. aeruginosa* have been extensively characterized [21–27]. These cell wall proteins are found in many other pathogenic bacteria and are classified according to amino acid sequence and function [28]. To date there are 11 *P. aeruginosa* lytic transglycosylases that have been described. Their functions range from cell division to aiding in the insertion of secretion systems and two component regulatory systems. They are attractive drug targets to enhance the activity of our most commonly used and safest antibiotics, the β-lactam class (penicillins, cephalosporins, carbapenems and monobactams).

Lts in general catalyze a cleavage reaction that breaks the glycosidic bond between the peptidoglycan building blocks, MurNAc and GlcNAc (**Figure 1**).

This reaction does not involve a water molecule but rather, an active site Glu or Asp residue functions as a general acid, donating a proton to the oxygen in the β-1,4 glycosidic linkage. Then the deprotonated active site residue acts a general base as a nucleophile to break the glycosidic bond. The result is a 1,6-anhydroMurNAc containing final peptide product. This unique cap on the muramyl peptide is a signal and a way for the cell wall peptidoglycan cleavage products to be trafficked for recycling [26]. The reaction shown in **Figure 1** is within the strand or "endolytic". Some Lts also catalyze an end of strand or "exolytic" cleavage.

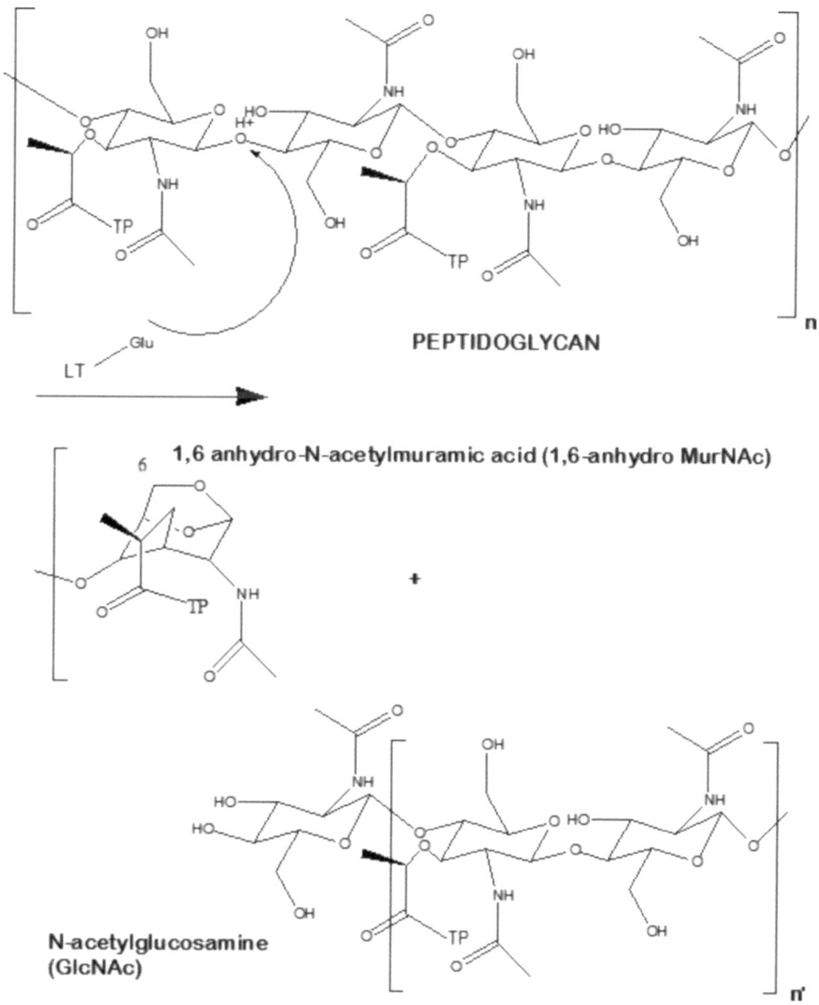

Figure 1.
Lt reaction in cell wall remodeling in Pseudomonas aeruginosa. When the transpeptidase (crosslinking function) of a PBP is inhibited by a β-lactam, the tranglycosylase function of the PBP continues to produce strands of uncrosslinked peptidoglycan (PG). The soluble Lt in the periplasm of Gram negative bacteria initiates recycling and cleavage of PG via endolytic (within strand) reaction. Once this first cleavage reaction occurs, the 1,6-anhydroMurNAC-GlcNAC containing fragments are cleaved and released. In P. aeruginosa, these 1,6-anhydromuramylpeptide fragments affect regulation of Amp C β-lactamase production. TP designates tetrapeptide.

Lts are classified according to amino acid motifs and function, into 6 distinct families. Even within a family, there is little sequence homology; however, the proteins in families do appear to share distinct folds (**Figure 2**). Lts are also divided into membrane (designated M in their nomenclature) and soluble (S) forms. It is hypothesized that these proteins are associated with numerous other cell wall proteins such as PBPs so that even the soluble Lts might be physically associated with the inner membrane of bacteria. Some Lts are also associated with the outer membrane, e.g., RlpA (see below) and likely have distinctive roles [29].

Lts serve many cellular functions including cell wall recycling, cellular division, insertion into cell wall of important structures like secretion systems and flagellar apparati. Lt redundancy is similar to that of the PBPs, and studies looking at gene knockouts of these proteins show that in *P. aeruginosa*, only loss of the RlpA LT is associated with a change in bacterial morphology [29]. Attempts to prepare multiple Lt knockouts were unsuccessful.

Recently significant research has been conducted on the Lts of *P. aeruginosa*, including structural and kinetic studies defining structure function relations in these varied proteins (reviewed in [26]). These studies are summarized next.

3.1 Kinetic studies of *P. aeruginosa* lytic transglycosylases

As previously indicated, *P. aeruginosa* possesses 11Lts: MltA, MltB/Slt35, MltD, MltF, MltF2, MltG, RlpA, Slt, SltB1 (SltB), SltB2 (SltG), and SltB3 (SltH).

In a tour-de-force of biochemical characterization, including synthesis, purification and characterization of the reaction of soluble forms of all 11 *P. aeruginosa* Lts with 4 synthetic substrates and *P. aeruginosa* sacculus to yield 31 distinct peptidoglycan (PG) products, Lee et al. [25] have thoroughly described the structure

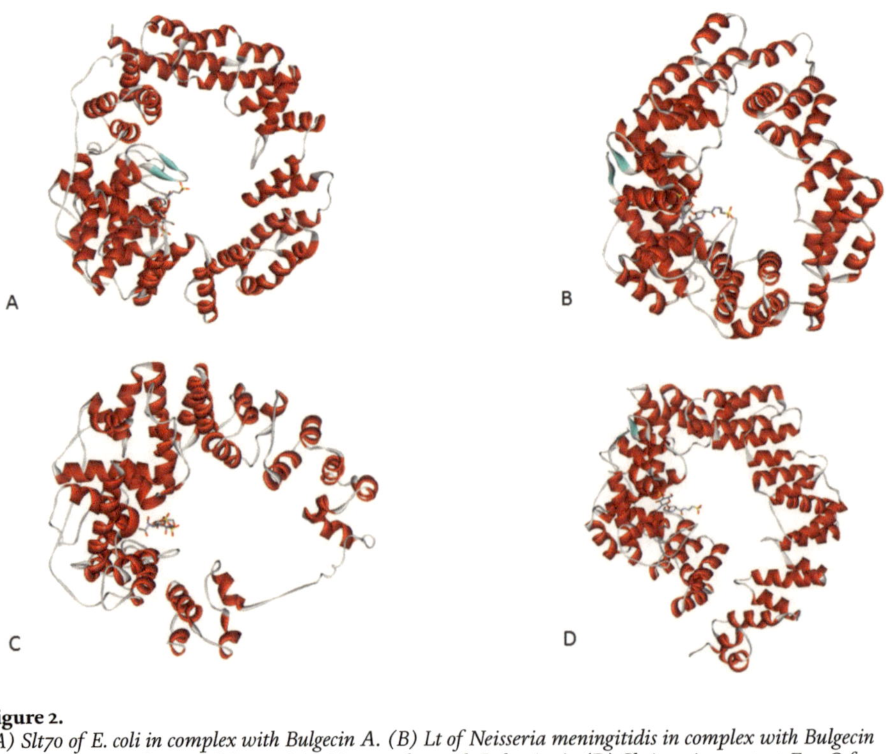

Figure 2.
(A) Slt70 of E. coli in complex with Bulgecin A. (B) Lt of Neisseria meningitidis in complex with Bulgecin A. (C) Lt Cj0843 of Campylobacter jejuni in complex with Bulgecin A. (D) Slt inactive mutant E503Q from Pseudomonas aeruginosa in complex with Bulgecin A.

function relationships for *P. aeruginosa* Lts. Of interest is that each solubilized Lt enzyme could perform both endolytic and exolytic reactions on the PG substrates.

Using the simplest synthetic substrate, NAG-NAM(pentapeptide)-NAG-NAM(pentapeptide), Lee et al. found that only MltB and the SltB1–3 Lts could recognize this substrate. A second substrate, a NAG-NAM(tetrapeptide)-NAG-anhydroNAM(tetrapeptide), incorporated the anhydroNAM that is likely recognized better by the exolytic Lts. For this substrate, MltA as well as MltB and the SltB1–3 Lts were able to react to convert 100% of the substrate to NAG-anhdroNAMtetrapeptide product. The soluble Lts, SltB1–3 of *P. aeruginosa* show the greatest activity in assays designed to study soluble proteins, as compared to solubilized membrane Lts [25]. These Lts were able to cleave NAG-NAM(pentapeptide)-NAG-NAM(pentapeptide) with specific activities of 0.4, 0.4 and 0.3 nanomoles of product/min/mg of protein respectively. Slt, the structural homolog of *E. coli* Slt70, showed no reaction with the synthetic peptidoglycan and slower turnover with the tetrapeptide substrate: 0.1 nmoles/min/mg.

3.2 Structural studies of the soluble Lts, Slt, SLtB1 and SltB3 of *P. aeruginosa*

X-ray crystal structures of Slt in its apo form as well as in complex with various synthetic PG substrates and reaction products demonstrated that this Lt has both exolytic and endolytic activity [23]. It is a donut shaped protein like Slt of *E. coli*. Notably, it is only after the binding of substrates that contain pentapeptide stems that it can exhibit endolytic activity due to a conformational change of the protein on substrate binding. A movie of this rearrangement is available in the supplementary material of reference [23]. Additional studies suggest protein–protein interactions with inner membrane PBPs are also important [26].

SltB1 [22] and SltB3 [24] have also been studied using x-ray crystallography. SLtB1 protein structures suggest that the protein forms a so-called "catenane" homodimeric structure in which the active sites face one another and are thus completely occluded. It is speculated that this soluble dimer may represent a form of activity regulation [22]. SltB3 is an exolytic enzyme with four distinct enzymatic domains within the donut shaped annular protein [24]. SltB3 can recognize PG substrates that are 4–20 sugars in length. These PG chains thread through the annular structure during catalysis.

3.3 Structural studies of the endolytic Lt, MltF

X-ray crystal structures of a solubilized MltF [21] show that this Lt binds a tetrapeptide stem of the substrate in an allosteric domain. Binding causes a large conformational change that leads to enzyme activation. In the kinetic studies, this solubilized membrane had very low activity with any of the 4 synthetic substrates or the *P. aeruginosa* sacculus. This raises some questions regarding the actual role of this Lt and whether the conformational changes are relevant when the protein is membrane bound.

4. Bulgecins as Lt inhibitors

Bulgecins were first described by Imato et al. in the 1980s [30, 31]. These natural analogs of GlcNAC-MurNAC are produced by various bacterial species including *Burkholderia mesoacidophila* and *Paraburkholderia acidophila* [32, 33], part of the *B. cepacia* complex. Bulgecins are produced together with sulfazecin, a monobactam antibiotic. Three different bulgecins are produced by these bacteria. Bulgecin A is produced in the highest amount and is the most active inhibitor of Lts (**Figure 3**, Bulgecin A).

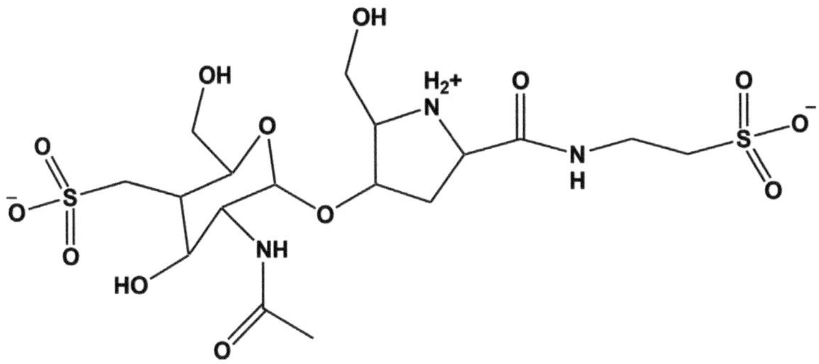

Figure 3.
Bulgecin A, the most active of the bulgecins of Paraburkholderia acidophila and Burkholderia mesoacidophila. The pyrrolidine ring (right side of the molecule) and the N-acetylglucosamine potion (GlcNAC) (left side other molecule) are features of Bulgecin A transition state structure.

Early research by Takeda Pharmaceuticals Japan led to the natural product isolation and purification of the bulgecins [30, 31]. It was discovered that when Bulgecin A was paired with a third generation cephalosporin, cefmenoxime, which targets PBP 3 of Enterobacteriaceae, large bulges were formed in the bacterial cell wall leading to osmotic lysis of the bacteria [30, 31]. Subsequently, investigators discovered the soluble Lt of *E. coli* and solved crystal structures of SltE in complex with Bulgecin A [34]. Through kinetic experiments, it was determined that Bulgecin A was a noncompetitive inhibitor of SltE with an IC_{50} of 0.5 μM. [35]. While Bulgecin A appeared to be a potent inhibitor of Lts in pathogenic Enterobacteriaceae and led to bacterial killing when paired with β-lactams affecting PBP3 particularly, development of the drug was halted for unknown reasons. Over the next decade, more advanced generation cephalosporins, as well as β-lactam-β-lactamase inhibitor combinations, carbapenems and fluoroquinolones were introduced into the clinic to address the growing problem of Gram negative resistance. Recently a natural product synthesis of the bulgecins was reported for the first time by Tomoshige et al. [36] prompting renewed interest in the use of Bulgecin A as an antimicrobial adjuvant, and possible drug optimization via medicinal chemistry.

Since the original discovery of the bulgecins and Slt in *E coli*, Lts have been characterized in many additional bacteria including *P. aeruginosa, Acinetobacter baumannii, Helicobacter pylori, Neisseria meningitides* and *Campylobacter jejuni* [21–25, 28, 34, 37–50]. In general, these organisms have many Lt enzymes with different functions, including endolytic (within strand) and exolytic (end of strand) cleavage of peptidoglycan. Many of these enzymes including those of *P. aeruginosa* have been expressed for biochemical assays of function, and inhibition by Bulgecin A. Crystal structures of many of the Lts of these organisms have also been obtained, some with substrates or Bulgecin A in the active site (**Figure 2**).

A recent publication shows that while *P. aeruginosa* possesses 11 known Lts, three appear to be the main targets of inhibition by Bulgecin A [27]. This work is discussed further below.

5. Microbiological effects of Bulgecin A

Bulgecin A in combination with cefmenoxime and other β-lactams has been studied against Enterobacteriaceae and reported in the original studies by Takeda Pharamceuticals [30, 31]. Later investigators studied Bulgecin A in combination with ampicillin in mouse models of *Helicobacter pylori* infection and found that the combination was effective in eradicating the organism, and Bulgecin A did not

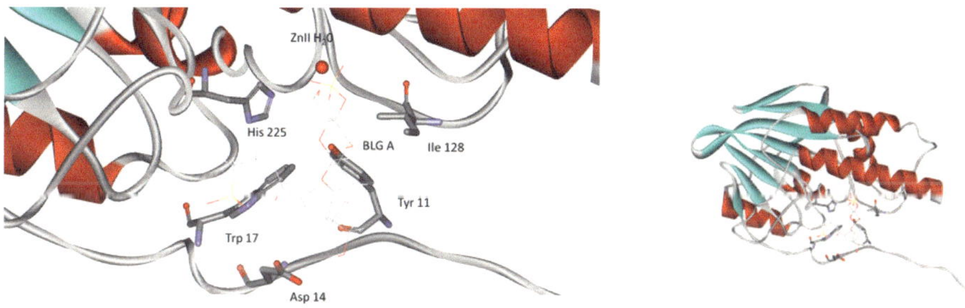

Figure 4.
L1 MBL (left) of Stenotrophomonas maltophilia with Bulgecin A; Bulgecin A sulphonates (yellow moieties, right) interacting with the ZnII site and with Asp 14 of the L1 protein.

appear to have specific toxicity in mice [51]. These investigators also studied Bulgecin A with *Neisseria gonorrhea* and *N. meningitides* strains that were resistant to penicillin and amoxicillin [48]. For strains with higher penicillin MICs not due to the presence of TEM-1 β-lactamase, Bulgecin A at concentrations of 19 mg/L, reduced the MICs from 0.5 to 0.09 mg/L for penicillin G, and 0.75 to 0.4 mg/L for amoxicillin.

Other investigators examined the effect of Bulgecin A as a metallo-β-lactamase inhibitor using L1 MBL of *Stenotrophomonas maltophila* as a model B1 (di Zn^{2+}) MBL enzyme (**Figure 4**). Simm et al. determined that the K_I for Bulgecin A was 150uM [52]. Later, our group investigated inhibition of VIM-1 using a Bulgecin A preparation from *B. mesoacidophila* and found that it also acted as an inhibitor of a second B1 MBL enzyme that is commonly found in *P. aeruginosa* in Europe, Asia and Canada, and rarely in the US [53].

Our group tested the Bulgecin A extracts from *B. mesoacidophila* against a variety of carbapenem resistant *P. aeruginosa* and *Acinetobacter baumannii* isolates with differing resistance mechanisms [53]. Although these were impure preparations, we found that small amounts were able to inhibit growth of these clinical isolates when combined with typical amounts of carbapenems to which the bacteria were otherwise resistant. The Bulgecin A-meropenem combinations proved effective whether carbapenem resistance was due to the presence of MBLs (VIM-1), hyperproduction of PDC (Amp C enzyme of *P. aeruginosa* in combination with OMP loss) or efflux. Tomoshige et al. using synthetic Bulgecin A were able to demonstrate bulge formation in *P. aeruginosa* PA01 as well as lysis in the presence of ceftazidime [36].

6. Slt, MltD and MltG are the main targets of Bulgecin A inhibition and potentiation of β-lactams that inhibit PBP2 and 3 in *P. aeruginosa*

Previously it was demonstrated that bulgecin A potentiated the bulge formation and lysis of *P. aeruginosa* in the presence of ceftazidime and meropenem [36] in a swarm assay [54]. Recently, Dik et al. [27] used individual transposon knockouts of. Lts in a susceptible *P. aeruginosa* strain, PA01 and further engineered a green fluorescent protein (GFP) gene into the bacteria. The various Lt knockout strains were exposed to ceftazidime, an inhibitor of PBP3 in *P. aeruginosa* and meropenem, an inhibitor of PBP2,3 and 4 [55] on agar medium containing propidium iodide. Bulge formation and bacterial cell lysis were monitored as a function of time by monitoring green fluorescence from viable cells, and red fluorescence during cell lysis, the red fluorescence arising from bacterial DNA interacting with the propidium iodide in the medium. In the presence of ceftazidime, the Slt and MltD knockouts formed bulges and showed lysis. The Slt knockout demonstrated significant bulge formation within 6 hours of exposure to ceftazidime, and lysis within

9 hours. Some of the other knockouts demonstrated minor bulge formation (MltA, MltG, MltF, SltB1, SltB3) at 9 hours but none showed cell lysis. The effect was even more dramatic in terms of rapidity of bulge formation and cell lysis when meropenem was used. In fact, this semi-qualitative assay that involves spotting the bacteria and β-lactam at a given distance onto agar had to be modified for meropenem, as lysis occurred too quickly compared to conditions for ceftazidime. In the case of meropenem, an inhibitor of PBP2, Slt showed the greatest bulge formation and lysis, followed by MltG.

The soluble forms of five of the Lt enzymes were purified and bulgecin A binding constants measured: Slt K_d = 8.5 ± 1.1 μM; MltD K_d = 1.4 ± 0.3 μM; MltG K_d = 24 ± 5 μM, SltB1 K_d = 160 ± 20 μM; RlpA K_d = 1200 ± 280 μM.

Dik et al. [27] also demonstrated via scanning electron microscopy that cell wall failure within the bulge is responsible for cell lysis, in the presence of β-lactams and Bulgecin A. Withdrawal of the β-lactam antibiotic leads to delayed recovery of cell morphology in the presence of Bulgecin A alone, suggesting further, the cooperative nature of the Lt and PBP functions in cell wall maintenance.

7. Future prospects for antibiotic enhancers to treat *P. aeruginosa* infections

Now that the syntheses of the bulgecins A, B and C have been accomplished and purified Lt enzymes of many bacteria are available with simple commercial high throughput assays such as ENZCHEK lysozyme™ (substrate is a fluorescein labelled sacculus from *Micrococcus lutei*), it should not be long before potent derivatives of bulgecins are developed through medicinal chemistry approaches. Combinations of enhancers with novel β-lactamase inhibitor/potent anti-pseudomonal β-lactams are possibilities in the future antibiotic arsenal.

8. Conclusions

Antibiotic resistance in *P. aeruginosa* is on the rise in both inpatient and outpatient settings. β-lactams remain among the most successful antibiotics due to their potency, efficacy and safety. Traditionally, β-lactamase inhibitors have proved able to extend the life of these valuable antibiotics. However, through a variety of resistance mechanisms, *P. aeruginosa* has eluded these approaches. Lts are novel cell wall enzymes that work in concert with PBPs to facilitate numerous cellular functions (insertion of secretion systems, cell division, etc.). When both Lts and PBPs are inhibited, bacteria exhibit abnormal bulging of cell wall and osmotic lysis. Bulgecins are naturally occurring compounds that inhibit Lts. Together with β-lactams, Bulgecin A can lead to effective bacterial killing, even when *P. aeruginosa* are resistant to the partner β-lactam antibiotics. Bulgecins are a novel β-lactam enhancer that may prove beneficial in the treatment of infections with resistant *P. aeruginosa*.

Acknowledgements

The author would like to acknowledge her affiliation and employment with the U.S. Department of Veteran Affairs.

Conflict of interest

The author is an employee of the U.S. Department of Veterans Affairs. The opinions expressed in this review are her own and do not reflect those of her employer.

Author details

Marion J. Skalweit
Louis Stokes Cleveland Veterans Affairs Medical Center and Case Western Reserve University School of Medicine, Cleveland, Ohio, USA

*Address all correspondence to: msh5@case.edu

IntechOpen

© 2019 The Author(s). Licensee IntechOpen. This chapter is distributed under the terms of the Creative Commons Attribution License (http://creativecommons.org/licenses/by/3.0), which permits unrestricted use, distribution, and reproduction in any medium, provided the original work is properly cited.

References

[1] Skalweit MJ. Profile of ceftolozane/tazobactam and its potential in the treatment of complicated intra-abdominal infections. Drug Design, Development and Therapy. 2015;**9**:2919-2925

[2] Chevalier S, Bouffartigues E, Bodilis J, Maillot O, Lesouhaitier O, Feuilloley MGJ, et al. Structure, function and regulation of *Pseudomonas aeruginosa* porins. FEMS Microbiology Reviews. 2017;**41**(5):698-722

[3] Richardot C, Plesiat P, Fournier D, Monlezun L, Broutin I, Llanes C. Carbapenem resistance in cystic fibrosis strains of *Pseudomonas aeruginosa* as a result of amino acid substitutions in porin OprD. International Journal of Antimicrobial Agents. 2015;**45**(5):529-532

[4] Cabot G, Zamorano L, Moya B, Juan C, Navas A, Blazquez J, et al. Evolution of *Pseudomonas aeruginosa* antimicrobial resistance and fitness under low and high mutation rates. Antimicrobial Agents and Chemotherapy. 2016;**60**(3):1767-1778

[5] Mesaros N, Nordmann P, Plesiat P, Roussel-Delvallez M, Van Eldere J, Glupczynski Y, et al. *Pseudomonas aeruginosa*: Resistance and therapeutic options at the turn of the new millennium. Clinical Microbiology and Infection: The Official Publication of the European Society of Clinical Microbiology and Infectious Diseases. 2007;**13**(6):560-578

[6] El Zowalaty ME, Al Thani AA, Webster TJ, El Zowalaty AE, Schweizer HP, Nasrallah GK, et al. *Pseudomonas aeruginosa*: Arsenal of resistance mechanisms, decades of changing resistance profiles, and future antimicrobial therapies. Future Microbiology. 2015;**10**(10):1683-1706

[7] Aeschlimann JR. The role of multidrug efflux pumps in the antibiotic resistance of *Pseudomonas aeruginosa* and other gram-negative bacteria. Insights from the Society of Infectious Diseases Pharmacists. Pharmacotherapy. 2003;**23**(7):916-924

[8] Guerin F, Isnard C, Cattoir V, Giard JC. Complex regulation pathways of AmpC-mediated β-lactam resistance in *Enterobacter cloacae* Complex. Antimicrobial Agents and Chemotherapy. 2015;**59**(12):7753-7761

[9] Potron A, Poirel L, Nordmann P. Emerging broad-spectrum resistance in *Pseudomonas aeruginosa* and *Acinetobacter baumannii*: Mechanisms and epidemiology. International Journal of Antimicrobial Agents. 2015;**45**(6):568-585

[10] Cavallari JF, Lamers RP, Scheurwater EM, Matos AL, Burrows LL. Changes to its peptidoglycan-remodeling enzyme repertoire modulate β-lactam resistance in *Pseudomonas aeruginosa*. Antimicrobial Agents and Chemotherapy. 2013;**57**(7):3078-3084

[11] Rodriguez-Martinez JM, Poirel L, Nordmann P. Extended-spectrum cephalosporinases in *Pseudomonas aeruginosa*. Antimicrobial Agents and Chemotherapy. 2009;**53**(5):1766-1771

[12] Karampatakis T, Antachopoulos C, Tsakris A, Roilides E. Molecular epidemiology of carbapenem-resistant *Pseudomonas aeruginosa* in an endemic area: Comparison with global data. European Journal of Clinical Microbiology & Infectious Diseases: Official Publication of the European Society of Clinical Microbiology. 2018;**37**(7):1211-1220

[13] Nguyen L, Garcia J, Gruenberg K, MacDougall C. Multidrug-resistant Pseudomonas infections: Hard to

treat, but hope on the horizon? Current Infectious Disease Reports. 2018;**20**(8):23

[14] Wright H, Bonomo RA, Paterson DL. New agents for the treatment of infections with gram-negative bacteria: Restoring the miracle or false dawn? Clinical Microbiology and Infection: The Official Publication of the European Society of Clinical Microbiology and Infectious Diseases. 2017;**23**(10):704-712

[15] Chen W, Zhang YM, Davies C. Penicillin-binding protein 3 is essential for growth of *Pseudomonas aeruginosa*. Antimicrobial Agents and Chemotherapy. 2017;**61**(1)

[16] Bassetti M, Vena A, Croxatto A, Righi E, Guery B. How to manage *Pseudomonas aeruginosa* infections. Drugs in context. 2018;**7**:212527

[17] Sader HS, Huband MD, Castanheira M, Flamm RK. *Pseudomonas aeruginosa* antimicrobial susceptibility results from four years (2012 to 2015) of the international network for optimal resistance monitoring program in the United States. Antimicrobial Agents and Chemotherapy. 2017;**61**(3)

[18] Righi E, Peri AM, Harris PN, Wailan AM, Liborio M, Lane SW, et al. Global prevalence of carbapenem resistance in neutropenic patients and association with mortality and carbapenem use: Systematic review and meta-analysis. The Journal of Antimicrobial Chemotherapy. 2017;**72**(3):668-677

[19] Barrasa-Villar JI, Aibar-Remon C, Prieto-Andres P, Mareca-Donate R, Moliner-Lahoz J. Impact on morbidity, mortality, and length of stay of hospital-acquired infections by resistant microorganisms. Clinical Infectious Diseases: An Official Publication of the Infectious Diseases Society of America. 2017;**65**(4):644-652

[20] Simner PJ, Johnson JK, Brasso WB, Anderson K, Lonsway DR, Pierce VM, et al. Multicenter evaluation of the modified carbapenem inactivation method and the Carba NP for detection of Carbapenemase producing *Pseudomonas aeruginosa* and *Acinetobacter baumannii*. Journal of Clinical Microbiology. 2018;**56**(1)

[21] Dominguez-Gil T, Lee M, Acebron-Avalos I, Mahasenan KV, Hesek D, Dik DA, et al. Activation by allostery in cell-wall remodeling by a modular membrane-bound lytic transglycosylase from *Pseudomonas aeruginosa*. Structure. 2016;**24**(10):1729-1741

[22] Dominguez-Gil T, Molina R, Dik DA, Spink E, Mobashery S, Hermoso JA. X-ray structure of catenated lytic transglycosylase SltB1. Biochemistry. 2017;**56**(48):6317-6320

[23] Lee M, Batuecas MT, Tomoshige S, Dominguez-Gil T, Mahasenan KV, Dik DA, et al. Exolytic and endolytic turnover of peptidoglycan by lytic transglycosylase Slt of *Pseudomonas aeruginosa*. Proceedings of the National Academy of Sciences of the United States of America. 2018;**115**(17):4393-4398

[24] Lee M, Dominguez-Gil T, Hesek D, Mahasenan KV, Lastochkin E, Hermoso JA, et al. Turnover of bacterial Cell Wall by SltB3, a multidomain lytic transglycosylase of *Pseudomonas aeruginosa*. ACS Chemical Biology. 2016;**11**(10):2936

[25] Lee M, Hesek D, Dik DA, Fishovitz J, Lastochkin E, Boggess B, et al. From genome to proteome to elucidation of reactions for all eleven known lytic Transglycosylases from *Pseudomonas aeruginosa*. Angewandte Chemie. 2017;**56**(10):2735-2739

[26] Dik DA, Fisher JF, Mobashery S. Cell-wall recycling of the gram-negative bacteria and the nexus to

antibiotic resistance. Chemical Reviews. 2018;**118**(12):5952-5984

[27] Dik DA, Madukoma CS, Tomoshige S, Kim C, Lastochkin E, Boggess WC, et al. Slt, MltD and MltG of *Pseudomonas aeruginosa* as targets of Bulgecin A in potentiation of beta-lactam antibiotics. ACS Chemical Biology. 2019. Feb 15;**14**(2):296-303. PubMed PMID: 30620575

[28] Dik DA, Marous DR, Fisher JF, Mobashery S. Lytic transglycosylases: Concinnity in concision of the bacterial cell wall. Critical Reviews in Biochemistry and Molecular Biology. 2017;**52**(5):503-542

[29] Jorgenson MA, Chen Y, Yahashiri A, Popham DL, Weiss DS. The bacterial septal ring protein RlpA is a lytic transglycosylase that contributes to rod shape and daughter cell separation in *Pseudomonas aeruginosa*. Molecular Microbiology. 2014;**93**(1):113-128

[30] Imada A, Kintaka K, Nakao M, Shinagawa S. Bulgecin, a bacterial metabolite which in concert with β-lactam antibiotics causes bulge formation. The Journal of antibiotics. 1982;**35**(10):1400-1403

[31] Nakao M, Yukishige K, Kondo M, Imada A. Novel morphological changes in gram-negative bacteria caused by combination of bulgecin and cefmenoxime. Antimicrobial Agents and Chemotherapy. 1986;**30**(3):414-417

[32] Horsman ME, Marous DR, Li R, Oliver RA, Byun B, Emrich SJ, et al. Whole-genome shotgun sequencing of two beta-proteobacterial species in search of the Bulgecin biosynthetic cluster. ACS Chemical Biology. 2017;**12**(10):2552-2557

[33] Loveridge EJ, Jones C, Bull MJ, Moody SC, Kahl MW, Khan Z, et al. Reclassification of the specialized metabolite producer Pseudomonas mesoacidophila ATCC 31433 as a member of the Burkholderia cepacia complex. Journal of Bacteriology. 2017;**199**(13)

[34] Thunnissen AM, Rozeboom HJ, Kalk KH, Dijkstra BW. Structure of the 70-kDa soluble lytic transglycosylase complexed with bulgecin A. Implications for the enzymatic mechanism. Biochemistry. 1995;**34**(39):12729-12737

[35] De pedro MA, Höltje JV, Löffelhardt W. Federation of European Microbiological Societies. Bacterial Growth and Lysis: Metabolism and Structure of the Bacterial Sacculus. New York: Plenum Press; 1993. pp. 241-244

[36] Tomoshige S, Dik DA, Akabane-Nakata M, Madukoma CS, Fisher JF, Shrout JD, et al. Total syntheses of Bulgecins A, B, and C and their bactericidal potentiation of the β-lactam antibiotics. ACS Infectious Diseases. 2018;**4**(6):860-867

[37] Artola-Recolons C, Carrasco-Lopez C, Llarrull LI, Kumarasiri M, Lastochkin E, Martinez de Ilarduya I, et al. high-resolution crystal structure of MltE, an outer membrane-anchored endolytic peptidoglycan lytic transglycosylase from *Escherichia coli*. Biochemistry. 2011;**50**(13):2384-2386

[38] Artola-Recolons C, Lee M, Bernardo-Garcia N, Blazquez B, Hesek D, Bartual SG, et al. Structure and cell wall cleavage by modular lytic transglycosylase MltC of *Escherichia coli*. ACS Chemical Biology. 2014;**9**(9):2058-2066

[39] Artola-Recolons C, Llarrull LI, Lastochkin E, Mobashery S, Hermoso JA. Crystallization and preliminary X-ray diffraction analysis of the lytic transglycosylase MltE from *Escherichia coli*. Acta Crystallographica.

Section F, Structural Biology and Crystallization Communications. 2011;**67**(Pt 1):161-163

[40] Byun B, Mahasenan KV, Dik DA, Marous DR, Speri E, Kumarasiri M, et al. Mechanism of the *Escherichia coli* MltE lytic transglycosylase, the cell-wall-penetrating enzyme for type VI secretion system assembly. Scientific Reports. 2018;**8**(1):4110

[41] Crepin S, Ottosen EN, Peters K, Smith SN, Himpsl SD, Vollmer W, et al. The lytic transglycosylase MltB connects membrane homeostasis and in vivo fitness of *Acinetobacter baumannii*. Molecular Microbiology. 8 Jun 2018. PubMed PMID: 29884996

[42] Fibriansah G, Gliubich FI, Thunnissen AM. On the mechanism of peptidoglycan binding and cleavage by the endo-specific lytic transglycosylase MltE from *Escherichia coli*. Biochemistry. 2012;**51**(45):9164-9177

[43] Lamers RP, Nguyen UT, Nguyen Y, Buensuceso RN, Burrows LL. Loss of membrane-bound lytic transglycosylases increases outer membrane permeability and β-lactam sensitivity in *Pseudomonas aeruginosa*. MicrobiologyOpen. 2015;**4**(6):879-895

[44] Schaub RE, Chan YA, Lee M, Hesek D, Mobashery S, Dillard JP. Lytic transglycosylases LtgA and LtgD perform distinct roles in remodeling, recycling and releasing peptidoglycan in *Neisseria gonorrhoeae*. Molecular Microbiology. 2016;**102**(5):865-881

[45] Suvorov M, Lee M, Hesek D, Boggess B, Mobashery S. Lytic transglycosylase MltB of *Escherichia coli* and its role in recycling of peptidoglycan strands of bacterial cell wall. Journal of the American Chemical Society. 2008;**130**(36):11878-11879

[46] Vijayaraghavan J, Kumar V, Krishnan NP, Kaufhold RT, Zeng X, Lin J, et al. Structural studies and molecular dynamics simulations suggest a processive mechanism of exolytic lytic transglycosylase from campylobacter jejuni. PLoS One. 2018;**13**(5):e0197136

[47] Williams AH, Wheeler R, Rateau L, Malosse C, Chamot-Rooke J, Haouz A, et al. A step-by-step in crystallo guide to bond cleavage and 1,6-anhydro-sugar product synthesis by a peptidoglycan-degrading lytic transglycosylase. The Journal of biological chemistry. 2018;**293**(16):6000-6010

[48] Williams AH, Wheeler R, Thiriau C, Haouz A, Taha MK, Boneca IG. Bulgecin A: The key to a broad-spectrum inhibitor that targets lytic transglycosylases. Antibiotics. 2017;**6**(1)

[49] Yin J, Sun Y, Sun Y, Yu Z, Qiu J, Gao H. Deletion of lytic transglycosylases increases B-lactam resistance in Shewanella oneidensis. Frontiers in Microbiology. 2018;**9**:13

[50] Zahrl D, Wagner M, Bischof K, Bayer M, Zavecz B, Beranek A, et al. Peptidoglycan degradation by specialized lytic transglycosylases associated with type III and type IV secretion systems. Microbiology. 2005;**151**(Pt 11):3455-3467

[51] Bonis M, Williams A, Guadagnini S, Werts C, Boneca IG. The effect of bulgecin A on peptidoglycan metabolism and physiology of *Helicobacter pylori*. Microbial Drug Resistance. 2012;**18**(3):230-239

[52] Simm AM, Loveridge EJ, Crosby J, Avison MB, Walsh TR, Bennett PM. Bulgecin A: A novel inhibitor of binuclear metallo-β-lactamases. The Biochemical Journal. 2005;**387**(Pt 3): 585-590

[53] Skalweit MJ, Li M. Bulgecin A as a β-lactam enhancer for carbapenem-resistant *Pseudomonas aeruginosa* and carbapenem-resistant *Acinetobacter*

baumannii clinical isolates containing various resistance mechanisms. Drug Design, Development and Therapy. 2016;**10**:3013-3020

[54] Anyan ME, Amiri A, Harvey CW, Tierra G, Morales-Soto N, Driscoll CM, et al. Type IV pili interactions promote intercellular association and moderate swarming of *Pseudomonas aeruginosa*. Proceedings of the National Academy of Sciences of the United States of America. 2014;**111**(50):18013-18018

[55] Davies TA, Shang W, Bush K, Flamm RK. Affinity of doripenem and comparators to penicillin-binding proteins in *Escherichia coli* and *Pseudomonas aeruginosa*. Antimicrobial Agents and Chemotherapy. 2008;**52**(4): 1510-1512

Chapter 6

Trends in Heavy Metals Tolerance and Uptake by *Pseudomonas aeruginosa*

Salah Abdelbary, Mamdouh S. Elgamal and Ayman Farrag

Abstract

Pseudomonas aeruginosa is considered as the most potent bacterial strain for solving heavy metals pollution problems. Pollution is the most of problems in our world which causing a lot of risks to human, animal, plant and ecosystem. Heavy metals pollution is an ever-increasing problem in developing nations. Release of heavy metals into the environment has increased in the recent years at an alarming rate. To remove heavy metals from environment, there are different methods such as physical, chemical and biological. The biological method includes microorganisms and plant which recorded high heavy metals removal, safe and low-cost method. Microorganisms remove heavy metals from environment by different mechanisms according to their types. Thus, microbes are used as potential candidates of bioremediation that can adapt quickly to the changing noxious environment and be utilized for toxic metal remediation. In bacterial uptake and tolerance to heavy metals, *Pseudomonas aeruginosa* recorded potential role in bioremediation of different heavy metals with high removal percentage comparison with other bacterial strains. Chapter discusses the roles and trends of *Pseudomonas aeruginosa* in heavy metals tolerance and uptake as potential bacterial strain.

Keywords: pollution, *Pseudomonas aeruginosa*, heavy metals, bioremediation, tolerance, plasmid and resistant genes

1. Introduction

Heavy metals are wide distribution elements in environment have a high density, atomic weight and multiple applications in different fields [1]. Heavy metal pollution causes environmental problems to human, animal, plant and ecosystem [2]. This pollution resulted in accumulation of different heavy metals in to soil and water. Different heavy metals such as copper, cadmium, nickel, lead, chrome and mercury, etc. released into natural sources and recorded high accumulation then have toxic effects on human health and biological system [3].

To remove of this heavy metals, there are different biotechnological methods were done such as bio-mineralization, bio-sorption, phyto-stabilization, hyper-accumulation, bio-stimulation, rhizo-remediation, myco-remediation, cyano-remediation and geno-remediation [4].

In nature, microorganisms play a potential role for a recycling and degradation of accumulated heavy metals which decrease their toxicity. This microorganisms as fungi, bacteria and algae were recognized [5]. Also, bioremediation of heavy metals by microorganisms is an economic and effective strategy because of its high efficiency, low cost and eco-friendly nature. Additionally, microbial bioremediation is done by interaction of microbe-metal for accumulation and detoxification of heavy metals [6].

Subsequently, some of bacterial strains can remove heavy metals from polluted soils as *Pseudomonas* sp., *Arthrobacter* sp., *Alcaligenes* sp., *Corynebacterium* sp., *Bacillus* sp., *Flavobacterium* sp., *Mycobacterium* sp., *Azotobacter* sp., *Rhodococcus* sp. and *Methanogens* [7]. Bacterial strains have different mechanisms to adapt and remove of heavy metals from polluted environments [8]. Different mechanisms are responsible for adapted of bacterial strains to grow at high concentrations from heavy metal. Also, mechanisms include accumulation and uptake, sorption of metal, enzymatic reduction or oxidation and extracellular precipitation. Additionally, metal tolerance was done by molecular mechanisms by having resistant genes [9]. Heavy metals resistant bacterial strains able to grow under exposing to high concentrations and have potentiality in bioremediation of high content of heavy metals in soils [10]. Finally, *Pseudomonas aeruginosa* tolerated and removal the highest dose of heavy this metals as compared with different bacterial strains [11].

2. Problems and risks of heavy metals on environment

2.1 Heavy metals pollution sources

Firstly, pollution of environment is a huge problem in the world because of increasing of industrial activities resulting the toxic compounds which lead to contamination of soils and clean water [12]. Toxic heavy metals pollution is a wide distribution throughout world countries along with progress of different industries which chromium, copper, nickel cadmium and mercury were observed as the most common heavy metals that widespread and used causing environmental pollution [13]. Increased industrial activity and demand for heavy metals like arsenic, nickel, chrome, copper, lead, mercury, manganese, zinc cadmium and many more has recorded increasing in amounts of heavy metals at polluted wastewater [14]. Additionally, metallurgical and mining industries are generated high amounts of heavy metals in wastewaters and accumulate in soil and water. [15]. Also, agricultural activities using pesticides, fertilizers and crop preservatives which participate in releasing of heavy metals in to the environment [16].

2.2 Heavy metals effects on living organisms

Heavy metals are any inorganic metallic compound that can exert their toxicity via binding to the thiol group of the enzyme and the disulfide bond that contribute the stability of the enzyme. The metals have high affinity to the disulfide bridge between two cysteine residues in any protein compound. Heavy metals are very dangerous to living organisms especially human as certain of them cause DNA damage and their carcinogenic effects in animals and man are probably causally related to their abilities to cause mutation [17]. Also, heavy metals divided into nonessential metals (lead, cadmium, mercury and nickel) and essential metals (zinc, copper, iron and manganese). Because of their high toxicity, lead and cadmium represented the major heavy metals pollutants. Cadmium is released to ecosystem during electroplating, effluents from textile, mine tailing, tannery, leather, and galvanizing industries (cadmium batteries) [18]. Due to their high prevalence as contaminants, heavy metals have an excellent concern to environmental problems in soluble type that area unit terribly venomous to biological systems causing cancer [19]. Also, heavy metal compounds in soluble forms might be harmful for living organisms of ecosystem by entering of the food chain [16].

In 2015, ATSDR prepared the priority list of hazardous substances and the results recorded that cadmium and lead were in the seventh position and second position, respectively. Also, The International Agency for Research on Cancer classified cadmium and cadmium compounds as a group (1) carcinogen and lead compounds as group (2) carcinogens [20]. Additionally, cadmium cause renal dysfunction especially in the proximal tubular cells which considered as the main site accumulation of cadmium. Also, cadmium cause demineralization of bone either indirectly, as a result of renal dysfunction or directly by the bones damaging. Furthermore, nickel has carcinogenic, neurotoxic, hemato-toxic, reproductive toxic, immune-toxic, genotoxic, nephrotoxic, pulmonary toxic and hepatotoxic effects [21]. On the other hand, mercury cause risks effects as neurotoxic substance because it accumulates and increase its content in food chains [22].

Subsequently, if plants exposure to excess copper concentrations, it generates oxidative stress causing damage to macromolecules and metabolic pathways disturbance. Also, excessive manganese accumulates in leaves and causing photosynthetic rate reduction [23]. Plants can uptake high concentration of heavy metals that present in causing adversely effect of symbiosis, the growth and consequently effects on the crops yields [24].

In addition, chromium (hexavalent form) is the most toxic chromium species which used in some industries as leather processing. Chromium is a carcinogenic substance especially in case of the lung and enter through inhalation. Chromium toxicity comes from its potentiality to cause allergic reactions and be corrosive [25].

Furthermore, heavy metals effect on microorganisms and induced physiological and morphological changes in microbial populations [26]. When microorganisms exposed to heavy metal stress, they produce antioxidant enzymes under toxic conditions and tolerance of this stress such as some resistant bacterial strains [27]. In naturally, microbes response to toxicity of heavy metals depends on high concentration and its resistance and tolerance mechanisms [28].

3. Bioremediation of heavy metals by microorganisms

Bioremediation meaning use of microbial metabolism to remove of pollutants. It can occur on its own and this called intrinsic bioremediation or can be done by addition of fertilizers to stimulate of microbial bioavailability inside medium and this called bio stimulation. Also, in some cases the addition of other microbial strains into medium to enhance the resident microbial population's and increase their ability to remove of heavy metals. Microorganisms that used to perform this function of bioremediation known as bio-remediators [29]. These microorganisms have developed unique resistance mechanisms which allowing to survive and remove high concentrations of heavy metals from environments [30]. Subsequently, bioremediation considered as alternative to chemical techniques by using microorganisms for biodegrading and detoxify of heavy metals from polluted soils and wasted groundwater [31].

In addition, biosorption defined as the use of biomass to remove heavy metals from environment by using microorganisms as (bacterial strains, fungal strains and algae) or plant extracts. It represents as a low-cost method and environmentally friendly for bioremediation of heavy metals and management of resource [32]. The need for an efficient and inexpensive method has interested in case of bio-sorption and bio-accumulation processes using microorganisms as profit systems for removing of heavy metal [14].

Fungal and bacterial strains have been reported to remove high concentrations of heavy metals from polluted environment using biosorption and bioaccumulation techniques [33]. Bioaccumulation is a process which involves two aspects; active metal uptake and passive metal uptake and may be carried out by any living organism with the ability to withstand the toxic effects of a particular metal ion [34]. Additionally, utilization of potential microbial populations in biosorption process to transform or adsorb heavy metals either by live and dead biomass or by their products have produced to for detoxify of heavy metals forms whether in particulates or as soluble form. Negative charged of microbial cell surface as a result of the presence of different functional groups such as hydroxyl, amines, carboxylic and phenolics give microorganisms an ability for binding with different cationic heavy metals [35].

As above, microbial strains have different mechanisms for reducing the toxicity of heavy metals through its intracellular and extracellular precipitation, binding of elements to cell wall, adsorption on polysaccharides or by export via various transporters [36]. Also, in wide variety of bacterial strains especially in genus "*Pseudomonas*" resistance to heavy metals, disinfectants, antibiotics, detergents and different toxic substances were observed. *Pseudomonas* considered as one of the most indicators bacterial strains for measuring contamination in environment [37, 38].

4. *Pseudomonas aeruginosa* as the most potent bacterial strain for tolerating and uptake heavy metals

4.1 Bacterial community and *Pseudomonas* classification

Bacteria are microorganism play important role in living world. It represents approximately 10^8 g of the total living world biomass. They used as bio sorbent because their ubiquity, small size, and ability to grow under different conditions such as *Pseudomonas*, *Bacillus*, *Escherichia*, *Micrococcus*, and *Streptomyces* species and used for bioremediation of heavy metals by using functional groups and metal chelating agents present on cell wall to make metal binding [35].

In addition, *Pseudomonas aeruginosa* classified as Gram negative bacteria, Gamma Proteobacteria, aerobic, rod and belonging to family *Pseudomonadaceae* which tolerate some heavy metals such as copper, chromium, cadmium and nickel [39]. Also, it is tolerant to different physical conditions and resistant to high concentrations from most of heavy metals, dyes, salts, weak antiseptics and antibiotics [40]. Several studies reported that *Pseudomonas aeruginosa* has efficiency for metal uptake which biosorption of cadmium (II) and lead (II) ions from solution using lyophilized *P. aeruginosa* (PAO1) cells were observed under different conditions [41].

4.2 Tolerance and resistance mechanisms by *Pseudomonas aeruginosa*

Pseudomonas aeruginosa has three different mechanisms for resistance of heavy metals: Firstly, accumulation of specific ion can be diminishing not by interference with uptake but by using of the heavy metal ion active extrusion from cells and this mechanism is only specific for *Pseudomonas aeruginosa*. Secondly, cations especially the "Sulfur lovers" can be segregated in to complex compounds by thiol-containing molecules and then ejected from the cell. Thirdly, some metal ions could also be reduced to a less deadly aerophilic state by the complicated enzymes and special oxidization mechanisms within the cells. Finally, for many metals, resistance and homoeostasis is a combination of two or three of the mentioned basic mechanisms that is the case which *Pseudomonas aeruginosa* success. *Pseudomonas aeruginosa* produce an extracellular compound with yellowish green fluorescence, called Pyoverdin, which functions as a byproduct. The production of Pyoverdin, formerly called fluorescein, is concomitant with the production of another byproduct, Pyochelin and produce other types of soluble pigments, the blue pigment pyocyanin [40].

Additionally, *Pseudomonas aeruginosa,* yet as different metal-tolerant bacterium, develop varied detoxification and/or tolerance mechanisms, such metal reduction, precipitation as metal salts, animate thing sequestration, binding to metallothioneins and therefore the removal of excessive metal ions out of the cell by transport (efflux pump). Removal of excessive metal ions out of the cell by flow pump is achieved by varied proteins driven by ATP chemical reaction (ATPases) and ion diffusion transporter that acts as chemiosmotic ion-proton money dealer and therefore the Resistance Nodulation Division (RND) transporters that mediate nucleon driven flow [42]. Many mechanisms are evolved to resist metal uptake. These embrace the discharge of metal outside the microbial cell, metal storage within the microbial cell and reduction of virulent metal to fewer virulent forms [43].

4.3 Calculation of removal percentage by *Pseudomonas aeruginosa*

Firstly, *Pseudomonas aeruginosa* has a good ability to resist and accumulate metal ions such as HgCl, $MgSO_4$, Zn_2O_3, $MgCO_3$, $CuCl_2$ and $CdCl_2$ [35]. In addition, study suggest that *Pseudomonas aeruginosa* can be an effective measure for heavy metals compensation and recorded best achieved with 15% metal concentration of copper and zinc, which showed a reduction in free ion concentration about 79.1% at 48 hours, respectively, 52.4% at 72 hours of an incubation. There was a biodegradable chromium of 41.6% at 72 hours of incubation with a 5% concentration of ion and with reduced concentrations of metal reduction of reduced Cr-ion. The reduction of the concentration of free metal ions was observed at 61.0% of the 10% solution after 24 hours of incubation [39]. In another study, *Pseudomonas aeruginosa* presents a potential sorbent for the removal of heavy metals contained in groundwater. The results of the experiments showed that these bacteria can break an average of 81% of heavy metals as low-cost, highly-efficient [44].

In addition, the results of the study show the potential for the isolated *Pseudomonas aeruginosa* (S7) which resistance to heavy metals in the treatment of heavy metals contaminated solutions. Further study investigates their ability to remove heavy metals in pollution area and genetic traits for tolerance to heavy metals were recorded [45].

Also, bacterial strains isolated from the drainage of Kakuri characterized and subjected to the salt concentration of various heavy metals and limited its ability to carry heavy metal and recorded minimally inhibitory concentrations (MIC). This demonstrates their ability to tolerate and live in an atmosphere with high metal salts. Eight (8) heavy metals were examined and included; $ZnSO_4$, $CdCl_2$, $CoCl_2$, $K_2Cr_2O_7$, $CuSO_4$, $HgCl_2$, $NiCl_2$ and $PbCl_2$ [46].

Furthermore, other study showed that 90.4% of mercury biosorption was observed on combinations of cultures *Bacillus subtilis* and *Pseudomonas aeruginosa* 78.5 and 99.3% respectively. Also, the time required for maximum sorption of maximum mercury amount is 40 and 60 minutes for a mixture of cultures and *Bacillus subtilis* and *Pseudomonas aeruginosa*. In addition, biosorption of chromium showed that 77.6% of cultures, 60.5 and 81.3 for *Pseudomonas aeruginosa* and *Bacillus subtilis* respectively. Also, Arsenic biosorption is carried out using the same biomass as described above by achieving sorption of 30, 32 and 28% of mixed cultures (*Pseudomonas aeruginosa* and *Bacillus subtilis*) [47].

In a test in which the removal of heavy metals from waste water is an important target, heavy metals biosorption on biomass of *Pseudomonas aeruginosa*, immobilized carbon activated on granular, it has been studied in batch and column systems. In this batch system, the adsorption of heavy metals is reached between 20 and 50 minutes, and the best dose of bio solids is 0.3 g/l. so, the efficiency of the biosorption is 84, 80, 79, 59 and 42% For Cr, Ni, Cu, Zn and Cd respectively [48].

In another study, *Pseudomonas aeruginosa* isolated from the waste water of electroplating industry, is able to absorb chromium, nickel and zinc, by 20% concentration. The highest percentage of the reduction was observed in nickel after 10 days and lowest for 10 days chromium, so the bacteria can be used as bio sorbents [49].

In addition, another study was investigated for biosorption of ionic cadmium by *P. aeruginosa* under varying conditions which the values have the first pH of the cadmium solution ranges from 1 to 7, the maximum removal of cadmium is obtained at pH 6. From the perspective of the application of the procedure, the time for bio sorbents was 70 minutes and biosorption concentrated (1 g/l) is a suitable bio sorbent for treatment of cadmium ion (up to 100 ppm) [50].

Additionally, it has been found that the adsorption of heavy metals by *Pseudomonas aeruginosa* bio flocculant is influenced by the first metal focus, the concentration of bio flocculant and the pH of the solution. The study showed that microbial potential bio flocculant has been used as a bio remedial tool in the treatment of contaminated wastewater with heavy metals [51].

4.4 Plasmid mediated heavy metals in *Pseudomonas aeruginosa*

Pseudomonas aeruginosa launches resistance to heavy metals such as cadmium, chromium, nickel and lead. DNA plasmid was isolated from *P. aeruginosa* and has been defined as pBC15 and the plasmid size is about 23 kb [52]. Also, results of heavy metal tolerance and accumulation experiments concluded that *Pseudomonas aeruginosa* bacterial strain has the tendency for tolerate heavy metals due to it has plasmid that carry genes and play important role in tolerance of heavy metals, so it will be promising for new trends in heavy metals bioremediation and bioaccumulation in the future [2].

In addition, genes are set for the degradation of environmental pollution, such as heavy metals, toluene, acids, and pesticides, Halogen and this toxic waste. So, plasmids are required for each compound. It is not that one plasmid reduces all toxic compounds from other groups [53].

In bioremediation of chromium, bacterial strains show chromosome plasmid resistance and reduced enzyme coordination. In molecular engineering, it can now extract stress by improving even under stress conditions [4]. Also, it has been reported that the plasmid resistance gene is determined in pathogenic bacteria of the genus *Escherichia*, *Salmonella*, *Shigella*, *Klebsiella*, *Aeromonas* and *Pseudomonas* which determining factors for resistance to heavy metals such as cadmium, cobalt, nickel, zinc, and mercury, also different groups of drugs, such as tetracycline, quinolones, aminoglycosides and β-lactam [54].

Additionally, *Pseudomonas aeruginosa* use different types of mechanisms in response to heavy metals stress. These mechanisms can be encoded with chromosomal genes, but more often resistance is located on plasmid [55]. Plasmid curing in *Pseudomonas aeruginosa* is a testament to the relationship between genetic presentation and the transmission of a specific feature in heavy metals tolerance and removal. Various approaches have been developed to cure of plasmid, including chemical and physical agents for the elimination of plasmid [56].

4.5 Evaluation of resistant genes in *Pseudomonas aeruginosa*

Firstly, metals-microbial interactions might have several environmental implications. Main resistance mechanisms for some heavy metals as (Cu^+, Zn^+ and Ni^{2+})

were active efflux transporters. Also, in bacterial strains, molecular basis of zinc resistance determined by presence of znt-related genes. In addition, it investigates adaptation of *Bacillus cereus* and has znt genes. Heavy metal resistant genes identified in *Pseudomonas aeruginosa* as CZC genes [57].

Subsequently, the ncc, czc, mer and chr genes responsible for heavy metals resistance to different heavy metals as Cr, Zn, Hg and Ni and which the genes have high homology to the chrB, czcD, mer and nccA genes [58].

In *Pseudomonas*, there are 6 genes in resistance of cadmium were identified formed from 3 gene clusters as cadA2R, czcCBA1 and colRS. The homologs of the first two gene clusters were predicted as metal efflux systems [59].

Finally, conjugative plasmid (pUM505) isolated from *Pseudomonas aeruginosa* possesses a putative (31.292 kb) mobile element in addition to possessing chr genes that confer chromate resistance to *Pseudomonas* contains two putative mer operons which could confer resistance of mercury. Furthermore, the Mpe contains genes related with the virulence of *Pseudomonas aeruginosa* [60].

5. Conclusion

Heavy metals pollution cause problems and effect on soil, water, plant, animal, human and ecosystem. Heavy metals cause health risks to human lead to cancer. Also, the highest removal percentage of heavy metals from environment is recorded by microorganisms and plants. Microbial community recorded high tolerance and uptake to different heavy metals such as bacteria, fungi and algae. *Pseudomonas aeruginosa* is the most potent bacterial strains which tolerating and removal heavy metals. Tolerance and removal of heavy metals occurred by different mechanisms. Additionally, *Pseudomonas aeruginosa* recorded high removal percentage from different heavy metals such as cadmium, nickel, lead, chromium, mercury, copper and zinc. Also, different studies insured that high removal recorded at optimum conditions for growth and biomass produced. Conditions included pH, temperatures, biomass dose and incubation periods. Finally, *Pseudomonas aeruginosa* can tolerate heavy metals using resistant genes and genes that carry on plasmid which play important role in increase efficacy of strain in bioaccumulation and tolerance. In future prospections, *Pseudomonas aeruginosa* will be promising in heavy metals bioremediation and bioaccumulation from environment and achieve high removal percentage after using genetic engineering and gene transfer.

Acknowledgements

Firstly, great thanks and appreciation to staff of Al-Azhar University, Faculty of Science for their supporting and encouragement.

Finally, I wish to thank my Mother, Brother, sisters, wife, sons (Khalid and Mohammed) and everyone in my family for their continual guidance.

Conflict of interest

The author declares no conflict of interest.

Abbreviations

Cd	cadmium
Cu	copper
Ni	nickel
Zn	zinc
Hg	mercury
Pb	lead
Mn	manganese
Cr	chromium
Kb	kilobase pairs
MIC	minimum inhibitory concentration

Author details

Salah Abdelbary*, Mamdouh S. Elgamal and Ayman Farrag
Botany and Microbiology Department, Faculty of Science, Al-Azhar University, Cairo, Egypt

*Address all correspondence to: salah.micro87@gmail.com

IntechOpen

© 2019 The Author(s). Licensee IntechOpen. This chapter is distributed under the terms of the Creative Commons Attribution License (http://creativecommons.org/licenses/by/3.0), which permits unrestricted use, distribution, and reproduction in any medium, provided the original work is properly cited.

References

[1] Tchounwou PB, Yedjou CG, Patlolla AK, Sutton DJ. Heavy metal toxicity and the environment. Molecular, Clinical and Environmental Toxicology. 2012;**101**:133-164

[2] Elgamal MS, Ahmed AF, Abdelbary S. Evaluation of nickel tolerance by identified Pseudomonas aeruginosa isolated from Egyptian polluted soils. Bioscience Research. 2018;**15**(1):518-529

[3] Dixit R. Bioremediation of heavy metals from soil and aquatic environment: An overview of principles and criteria of fundamental processes. Sustainability. 2015;**7**(2):2189-2212

[4] Samantaray D, Mohapatra S, Mishra BB. Microbial Bioremediation of Industrial Effluents. In Microbial Biodegradation and Bioremediation. Elsevier; 1 Jan 2014. pp. 325-339

[5] Kumar DMC. Biotechnological Advances in Bioremediation of Heavy Metals Contaminated Ecosystems: An Overview with Special Reference to Phytoremediation; 2013

[6] Rajendran P, Muthukrishnan J, Gunasekaran P. Microbes in Heavy Metal Remediation. Vol. 412003. pp. 935-944

[7] Girma G. Microbial bioremediation of some heavy metals in soils: An updated review. Journal of Resources Development and Management. 2015;**10**:62-74

[8] Access O. Isolation and identification of heavy metals tolerant bacteria from industrial and agricultural areas in Mauritius. Current Research in Microbiology and Biotechnology. 2013;**1**(3):119-123

[9] Hrynkiewicz K, Baum C. Application of microorganisms in bioremediation of environment from heavy metals. In: Environmental Deterioration and Human Health book. Springer; 2014:215-227

[10] Issazadeh K, Jahanpour N, Pourghorbanali F, Raeisi G. Heavy metals resistance by bacterial strains. Annals of Biological Research. 2013;**4**(2):60-63

[11] Maitra S. Study of genetic determinants of nickel and cadmium resistance in bacteria-A review. International Journal of Current Microbiology and Applied Sciences. 2016;**5**(11):459-471

[12] Kulshreshtha A, Agrawal R, Barar M, Saxena S. A review on bioremediation of heavy metals in contaminated water. IOSR Journal of Environmental Science, Toxicology and Food Technology. 2014;**8**(7):44-50

[13] Kumar RR. Isolation and molecular identification of metal tolerant bacteria and its heavy metal removal capacity. International Journal of Microbiology, Biochemistry and Molecular Biology. 2014;**14**(5):4-9

[14] Mathew BB, Krishnamurthy NB, Krishnamurthy TP. Bioaccumulation of heavy metals by fungi. International Journal of Environmental Chemistry and Chromatography. 2015;**1**(1): 15-21. Available from: https://www.researchgate.net/publication/305778861_BIOACCUMULATION_OF_HEAVY_METALS_BY_FUNGI

[15] Pirilä M. Adsorption and Photocatalysis in Water Treatment: Active, Abundant and Inexpensive Materials and Methods; 2015

[16] Gunasekaran B, Hakimi M, Kasim M, Salvamani S, Shukor MY. Field trials on heavy metals using alpha-chymotryopsin enzyme assay. Journal of Environmental. 2014;**2**(1):25-34

[17] Fazli MM, Soleimani N, Mehrasbi M, Darabian S, Mohammadi J, Ramazani A. Environmental health highly cadmium tolerant fungi: Their tolerance and removal potential. Journal of Environmental Health Science and Engineering. 2015;**13**:19

[18] Joo JH, Hussein KA. Heavy metal tolerance of fungi isolated from contaminated soil. Korean Journal of Soil Science and Fertilizer. 2012;**45**(4):565-571

[19] Moya M. Heavy metal tolerance of filamentous fungi isolated from polluted sites in Tangier, Morocco. African Journal of Microbiology Research. 2014;**3**(2):35-48

[20] Sowmya M, Hatha AAM. Cadmium and Lead Tolerance Mechanisms in Bacteria and the Role of Halotolerant and Moderately Halophilic Bacteria in their Remediation; 2017

[21] Bhagat N, Vermani M, Bajwa HS. Characterization of heavy metal (cadmium and nickle) tolerant gram negative enteric bacteria from polluted Yamuna River, Delhi. African Journal of Microbiology Research. 2016;**10**(5):127-137

[22] Cabral L, Yu RQ, Crane S, Giovanella P, Barkay T, Camargo FAO. Methylmercury degradation by *Pseudomonas putida* V1. Ecotoxicology and Environmental Safety. 2016;**130**:37-42

[23] Gill M, Gill M. Heavy metal stress in plants: A review. International Journal of Advanced Research. 2014;**2**(6):1043-1055

[24] Domisch T. Soil Biology. Vol. 322000. p. 1475

[25] Haddad HH. The effect of heavy metals cadmium, chromium and iron accumulation in human eyes. American Journal of Analytical Chemistry. 2012;**2012**:710-713

[26] Hussein KA, Joo JH. Heavy metal resistance of bacteria and its impact on the production of antioxidant enzymes. African Journal of Microbiology Research. 2013;7(20):2288-2296

[27] Ezzouhri L, Castro E, Moya M, Espinola F, Lairini K. Heavy metal tolerance of filamentous fungi isolated from polluted sites in Tangier, Morocco. African Journal of Microbiology Research. 2009;**3**(2):35-48

[28] Ropek V, Para A. The effect of heavy metal ions and their complexions upon growth, sporulation and pathogenicity of the Entomopathogenic fungus *Paecilomyces farinosus*. Polish Journal of Environmental Studies. 2003;**12**(2):227-230

[29] Sharma S. Bioremediation: Features, strategies and applications. The Asian Journal of Pharmacy and Life Science. 2012;**2**(2):202-213

[30] Monachese M, Burton JP, Reid G. Bioremediation and tolerance of humans to heavy metals through microbial processes: A potential role for probiotics. Applied and Environmental Microbiology. 2012;**78**(18):6397-6404

[31] Letters RB. Isolation of heavy metal resistant bacterial strains from the battery manufactured polluted environment. Romanian Biotechnological Letters. 2011;**16**(6):102-106

[32] Karman SB, Diah SZM, Gebeshuber IC. Raw materials synthesis from heavy metal industry effluents with bioremediation and phytomining: A biomimetic resource management approach. Advances in Materials Science and Engineering. 2015;1-21

[33] Kumar R, Dhir B. Potential of some fungal and bacterial species in bioremediation of heavy metals. 2014;**1**(2):213-223

[34] Brady D. Bioaccumulation of metal cations by yeast and yeast cell components [thesis]1992. p. 328. Available from: https://core.ac.uk/download/pdf/11985459.pdf

[35] Srivastava S, Agrawal SB, Mondal MK. A review on progress of heavy metal removal using adsorbents of microbial and plant origin. Environmental Science and Pollution Research. 2015;**22**(20):15386-15415

[36] Jarosławiecka A, Piotrowska-seget Z. Lead resistance in micro-organisms. Microbiology. 2014;**160**:12-25

[37] Hassan SHA, Abskharon RNN, Gad El-Rab SMF, Shoreit AAM. Isolation, characterization of heavy metal resistant strain of *Pseudomonas aeruginosa* isolated from polluted sites in Assiut city, Egypt. Journal of Basic Microbiology. 2008;**48**(3):168-176

[38] Technol JMB. Microbial & biochemical technology detection, identification and characterization of some heavy metals. Journal of Microbial & Biochemical Technology. 2016;**8**(3):226-230

[39] Awasthi G, Chester A, Chaturvedi R, Prakash J. Study on role of *Pseudomonas aeruginosa* on heavy metal bioremediation. International Journal of Pure & Applied Bioscience. 2015;**3**(4):92-100

[40] Abdul-sada HK. A resistance study of *Pseudomonas aeruginosa* to heavy metals. Journal of Veterinary Research. 2009;**8**(2)

[41] Konig-Péter A, Kocsis B, Kilár F, Pernyeszi T. Bio-adsorption characteristics of *Pseudomonas aeruginosa* PAO1. Journal of the Serbian Chemical Society. 2014;**79**(4):495-508

[42] Mihdir A, Assaeedi A, Abulreesh H, Osman G. Detection of heavy metal resistance genes in an environmental *Pseudomonas aeruginosa* isolate. British Microbiology Research Journal. 2016;**17**(1):1-9

[43] Naz T. Biosorption of heavy metals by *Pseudomonas* species isolated from sugar industry. Toxicology and Industrial Health. 2016;**32**(9):1619-1627

[44] Abbas A, Mohamed M, Zahir A. Biosorption of Heavy Metals by Pseudomonas Bacteria. IRJET; 2016. pp. 1446-1450. https://pdfs.semanticscholar.org/3d30/aece39605438c6c404f316b86a62aaa62578.pdf

[45] Mihdhir AA, Assaeedi AS, Abulreesh HH, Osman GE. Detection, identification and characterization of some heavy metals tolerant bacteria. Journal of Microbial and Biochemical Technology. 2016;**8**(8):226-230

[46] A.A H, K.K K, I A, Y M, E.E O. Evaluation of heavy metal tolerance level (MIC) and bioremediation potentials of *Pseudomonas aeruginosa* isolated from Makera-Kakuri industrial drain in Kaduna, Nigeria. European Journal of Experimental Biology. 2017;7(5):3-6

[47] Tarangini K. Biosorption of heavy metals using individual and mixed cultures of *Pseudomonas aeruginosa* and *Bacillus subtilis*. Defence Life Science Journal. 2009;**2**(4):442-447

[48] Orhan Y, Hrenovič J, Büyükgüngör H. Biosorption of heavy metals from wastewater by biosolids. Engineering in Life Sciences. 2006;**6**(4):399-402

[49] Pandian K, Thatheyus A, Ramya D. Bioremoval of chromium, nickel and zinc in electroplating effluent by *Pseudomonas aeruginosa*. Open Journal of Water Pollution and Treatment. 2014;**2014**(2):75-82

[50] Zeng X, Liu X, Jiang P, Li W, Tang J. Cadmium removal by *Pseudomonas aeruginosa* E1. In: 2009 International Conference on Energy

and Environmental Technology; ICEET 2009. Vol. 2. 2009. pp. 468-471

[51] Gomaa EZ. Production and characteristics of a heavy metals removing bioflocculant produced by *Pseudomonas aeruginosa*. Polish Journal of Microbiology. 2012;**61**(4):281-289

[52] Raja CE, Selvam GS. Plasmid profile and curing analysis of *Pseudomonas aeruginosa* as metal resistant. International Journal of Environmental Science and Technology. 2009;**6**(2):259-266

[53] Mishra A, Malik A. Recent advances in microbial metal bioaccumulation. Critical reviews in environmental science and technology. 1 Apr 2013;**43**(11):1162-222

[54] Ghaima KK, Mohamed AI. Resistance and bioadsorption of cadmium by *Pseudomonas aeruginosa* isolated from agricultural soil. International Journal of Applied Environmental Sciences. 2017;**12**(9):1649-1660

[55] Abdelatey LM, Khalil WK, Ali TH, Mahrous KF. Heavy metal resistance and gene expression analysis of metal resistance genes in gram-positive and gram-negative bacteria present in egyptian soils. Journal of applied sciences in environmental sanitation. 2011;**6**(2):201-211

[56] Irc A, Irc A, Irc C. Detection of plasmids and curing analysis in copper resistant bacteria. Biodiversitas Journal of Biological Diversity. 2016;**17**(1):296-300

[57] Joonu J, Averal HI. Heavy metal resistant CZC genes identification in *Bacillus cereus*, *Enterobacter asburiae* and *Pseudomonas aeruginosa* isolated from BHEL industry, Tamilnadu. Research & Reviews: Journal of Microbiology and Biotechnology. 2016;**5**(4):27-31

[58] Van Berkum P, Angle JS. Heavy metal resistance and genotypic analysis of metal resistance genes in gram-positive and gram-negative bacteria present in Ni-rich serpentine soil and in the rhizosphere of *Alyssum murale*. Chemosphere. 2007;**68**:360-367

[59] Nowicki EM, O'Brien JP, Brodbelt JS, Trent MS. Extracellular zinc induces phosphoethanolamine addition to *Pseudomonas aeruginosa* lipid A via the ColRS two-component system. Molecular Microbiology. 2015;**97**(1):166-178

[60] Hernández-Ramírez KC, Reyes-Gallegos R, Chávez-Jacobo V, Díaz-Magaña A, Meza-Carmen V, Ramírez-Díaz M. A plasmid-encoded mobile genetic element from *Pseudomonas aeruginosa* that confers heavy metal resistance and virulence. Plasmid. 2018;**98**:15-21

Chapter 7

The Beneficial Roles of *Pseudomonas* in Medicine, Industries, and Environment: A Review

Orji Frank Anayo, Onyemali Chidi Peter, Ukaegbu Gray Nneji, Ajunwa Obinna, Ezeanyanso Chika Scholastica and Lawal Oluwabusola Mistura

Abstract

This chapter intends to consider *Pseudomonas* versatility as regards the beneficial uses of *Pseudomonas* for production of primary metabolites including enzymes. This chapter will consider the use of *Pseudomonas* for production of secondary metabolites from various pathways which are equally useful in the industries and in medicine. *Pseudomonas* pigments and its usefulness in bioelectricity, medicine, etc. will equally be considered in this chapter. The authors are to integrate knowledge in versatility of *Pseudomonas* for use as agents of microbial production of biosurfactants for environmental cleanup, restoration, and remediation.

Keywords: pigments, enzymes, bioelectrochemical, bioremediation, medicine, industries

1. Introduction

Pseudomonas happens to be one of the most studied genera of prokaryotes (bacteria) and was first identified by Migula in 1894 at the end of the nineteenth century and described as Gram-negative, polar-flagellated, and rod-shaped bacteria. Thereafter the time of identification, the description of *Pseudomonas* has broadened; new methods have been developed to enhance the comprehensive study of the morphology and physiology of *Pseudomonas*. It is worthy to mention that the morphological features of these bacteria are common to several bacteria genera and are of little value in the positive identification of members of the genus (*Pseudomonas*). There are advanced nucleic acid-based methods that easily differentiate these bacteria from other similar genera, thus revealing the taxonomic relationships among various bacterial species including the genera *Pseudomonas*.

Pseudomonas is known to occupy a wide range of niches due to metabolic and physiological array. Their diversity enables *Pseudomonas* to adjust to exigent

IntechOpen

environmental conditions and withstand unfavorable conditions caused as a result of living and nonliving factors like oxygen, moisture, high and low temperature, etc.

The diversity of *Pseudomonas* determines wide research interest in this genus (*Pseudomonas*). *Pseudomonas aeruginosa* is ubiquitous in the environment (soil and water) and flourishes in individual with weakened or compromised immune system. The increasing resistance to multiple antibiotics makes these bacteria (*Pseudomonas aeruginosa*) virtually intractable once it colonizes the human host, and devastating effects can manifest.

Other species of *Pseudomonas aeruginosa* have the ability to degrade a large number of compounds that are recalcitrant to other bacterial species, thus producing secondary metabolites and biopolymers, making these strains useful in medicine, industries, and environment [1].

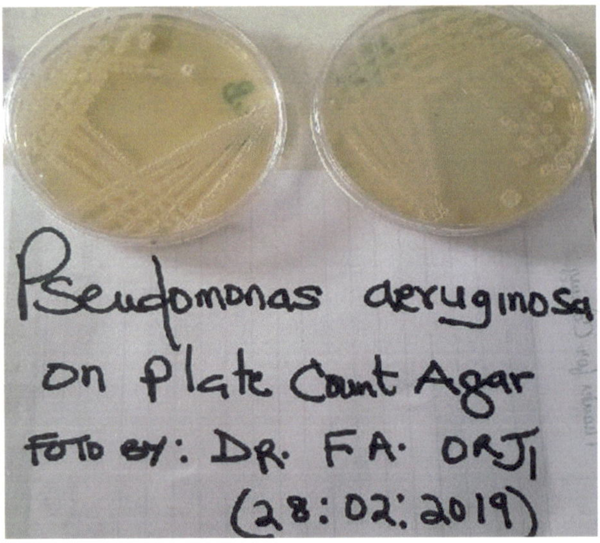

2. Beneficial roles of *Pseudomonas* species in medicine

Pseudomonas aeruginosa produces a wide range of compounds with bacteriostatic or bactericide activity, which are vital in the control of multiple drug-resistant (MDR) bacteria [2]. These compounds result from secondary metabolism and are referred to as secondary metabolites from various pathways including the polyketide and shikimic-chorismic acid pathways [3]. Studies have shown substances with antibacterial or antifungal activity secreted in the secondary metabolism of microorganisms could be applied in the management of human, animal, and plant diseases [4–6].

Pseudomonas aeruginosa strains produce compounds with antimicrobial properties, which include a group of peptides called pyocins and other heterocyclic compounds such as quinolines, phenylpyrroles, and phenazines [7]. These heterocyclic compounds eliminate microorganisms via DNA damage to cell depolarization in aiding the colonization of *Pseudomonas aeruginosa* [8].

Recent research has shown strain of *Pseudomonas aeruginosa* secreting an organometallic compound with inhibitory activity against MDR bacteria, including carbapenemase-producing *Klebsiella pneumoniae* and methicillin-resistant *Staphylococcus aureus* [9, 10]. Additionally, the same strain produced a phenazine-1-carboxylic acid (in low quantity) showing antifungal potential against *Botrytis cinerea* [11, 12].

2.1 Hydroxytyrosol from *Pseudomonas* for management of cardiovascular disease

Hydroxytyrosol, a phenylethanoid, is a molecule that has attracted high interests from the pharmaceutical industry due to its antimicrobial, anti-inflammatory, neuroprotection, antitumor, and chemomodulation effects and its role against cardiovascular diseases and metabolic syndrome. Interest in this molecule has led to a wide research on its biological activities, its valuable effects in humans, and how to synthetize new molecules from hydroxytyrosol. *Pseudomonas aeruginosa* has been reported to produce high yield of hydroxytyrosol by tyrosol into hydroxytyrosol via the immobilization of *Pseudomonas aeruginosa* resting cells in calcium alginate beads. The bioconversion yield reached 86% in the availability of 5 g L^{-1} of tyrosol when cells immobilized in alginate beads were carried out in single batches [13].

In addition, *Pseudomonas aeruginosa* strain 1001 produces an esterase (EstA) that has the ability to hydrolyse the racemic methyl ester of β-acetylthioisobutyrate to produce the (D)-enantiomer, which serves as a precursor of captopril. Captopril is an important drug in the management of congestive heart failure, hypertension, and diabetic nephropathy. This is achieved via the inhibition of angiotensin-converting enzyme [14].

3. Beneficial roles of *Pseudomonas* species in the industries

Pseudomonas aeruginosa has beneficial uses in various industrial and commercial sectors around the globe. These include waste degradation, oil refineries, textile products, agriculture, pulp and paper, mining, and explosive industries. They can also be used in commercial and household drain cleaners and degreasers, septic tank additives, general cleaning products, and odor control products.

Strains of *Pseudomonas aeruginosa* have been identified to play a vital role in the industries in the production of various compounds such as:

- Vanillin
- Rhamnolipids
- Protease
- Lipase
- Biopigments etc.

3.1 Vanillin synthesized by *Pseudomonas aeruginosa* for industrial application

It is one of the most important components of natural flavors; vanillin is broadly used in food, cosmetic, and pharmaceutical industries. Recent research reports the production of vanillin through microbial transformation using isoeugenol as a precursor by a novel strain of *Pseudomonas aeruginosa* ISPC2 isolated from the soil [15].

3.2 Rhamnolipids of *Pseudomonas aeruginosa* for industrial applications

Rhamnolipids are a class of glycolipid produced by *Pseudomonas aeruginosa*, among other organisms. They have a glycosyl head group, in this case a rhamnose moiety, and a 3-(hydroxyalkanoyloxy)alkanoic acid (HAA) fatty acid tail, such as

3-hydroxydecanoic acid [16–20]. Presently, rhamnolipids produced by *Pseudomonas aeruginosa* are the most essential class of biosurfactants, since the US Environmental Protection Agency has permitted their use in food products and other industrials applications. Rhamnolipids have been exploited by food industry particularly in increasing of food shelf life, considering to their high antimicrobial activity and physicochemical properties and also for emulsion stabilization, extends shelf life and inhibits hemophilic spores in Ultra-high temperature processing (UHT) soy-milk In combination with niacin [21]. A combination of niacin with rhamnolipids in salad extends its shelf life and inhibits the growth of mold. The shelf life in cottage cheese has been extended through the inhibition of mold and bacterial growth, especially Gram-positive and spore-forming bacteria. This has been achieved using a mixture of natamycin, nisin, and rhamnolipids. Consequently, these molecules and their mixtures are used to either eradicate food contamination directly, as food additive, or indirectly, as detergent formulation for cleaning surfaces that come in contact with the foods. Some other applications of rhamnolipids in the food industry especially in baking and confectioneries include promoting their dough or batter stability, volume and shape, structure, dough texture, width of the cut, and microbiological conservation [22]. In addition, they have been used to improve the properties of butter cream, nondairy cream filling for croissants, decoration cream and Danish pastries, and other fresh or frozen fine confectionery products [23]. Important food flavors are also synthesized from rhamnose which are gotten from rhamnolipids [24].

Furthermore, beyond the roles of rhamnolipids as agents that reduce surface and interfacial tension, they also have several other functions in food where they promote texture and shelf life of starch-containing products; regulate the agglomeration of fat globules; stabilize aerated systems; modify rheological properties of wheat dough; improve stability, consistency, and texture of oils and fat-based products; and inhibit separation. They aid in the general mixing of ingredients and can also retard the growth of molds and some bacteria in food. In ice cream and bakery formulations, rhamnolipids are also used to control consistency, retard staling, solubilize flavor oils, stabilize fats, and reduce spattering [25]. It has been demonstrated that rhamnolipids can be explored to control the attachment and to disrupt biofilms of individual and mixed cultures of the food-borne pathogens [26].

3.3 Industrial application of proteases from *Pseudomonas aeruginosa*

Pseudomonas aeruginosa is known to produce enzyme (proteases) with enzyme commission number (EC 3.4.21-24 and 3.4.99; peptidyl 3.4.21-24 and 3.4.99; peptidyl peptide hydrolases) which are applicable in industries that break down amino acids through the addition of water across peptide bonds and catalyze peptide synthesis in organic solvents and in solvents with low water content.

Proteolytic enzymes are largely found in all living organisms and are necessary for the growth of cells and cell differentiation [27]. *Pseudomonas aeruginosa discharges* extracellular protease making use of maltose as a major carbon source for the production, which has a maximum protease activity at pH 9.5, temperature of 37°C, and incubation time of 48 h [28].

3.4 Lipase from *Pseudomonas aeruginosa* for industrial applications

Lipases are glycerol ester hydrolases (EC 3.1.1.3) that hydrolyze ester linkages of glycerides at water-oil interface. Recent research shows that *Pseudomonas aeruginosa* are one of the best producers of lipase enzyme which have shown great potential with regard to their application in different sectors and industries [29]. The importance of microbial lipase production has increased in the last decades. Owing to their molecular structure and catalytic properties, these enzymes have specific biotechnological applications in various industrial sectors such as in cosmetology, environmental waste management, and foods, and specifically with regard to lipase, it acts as catalyst for synthesis of esters and for transesterification of the oil for the production of biodiesel [30].

3.5 Biopigments from *Pseudomonas aeruginosa* for industrial applications

The production of pigments by *Pseudomonas aeruginosa* strains as an area of research has been advanced over the years by different scholars. The various types of pigments produced by *Pseudomonas aeruginosa* are largely classed under the chemical name of phenazines. Phenazine compounds are of good biotechnological value. Apart from their possible uses as biopigments and coloring agents, they also function chemically as effective redox agents and can possess antimicrobial potentials. In current research, these redox-active agents have been heavily assessed for their potentials in acting as mediator compounds in bioelectrochemical applications. These bioelectrochemical applications can be channeled into the direct generation of electrical energy via instant electron harvesting from the cell surface of bacteria, as well as in the bioelectrochemically assisted remediation of harmful pesticides and chemicals and the potentials in the bioelectrochemically assisted fermentative production of certain commodity chemicals like butanol, ethanol, acetate, hydrogen gas, hydrogen peroxide, distilled water, and so on [31].

Phenazine pigments are naturally occurring heterocyclic compounds with varied chemical formula and pigmentations and are secreted mostly by *Pseudomonas* species. A good number of different structurally related phenazine compounds have being identified so far, totaling over 100. However, prominent among these phenazine pigments include chemical and color variants such as pyocyanin, pyoverdin, pyorubin, and pyomelanin. Most times, these pigments are produced by *Pseudomonas aeruginosa*; however a number of other *Pseudomonas* species produce similar chemical variants of these pigments [32]. A well-studied pigment for industrial application is pyocyanin. Pyocyanin with the chemical formula N-methyl-1-hydroxyphenazine is the pigment responsible for characteristic blue-green coloration of some *Pseudomonas* spp. It is a major molecule responsible for quorum sensing signaling for *Pseudomonas.* In Pharmacology, there are possible effects that pyocyanin can have on prokaryotic cells; such effects are linked to antimicrobiosis, and there is a related similarity in the chemical structure of flavins and flavoproteins as well as flavins adenine mono-/dinucleotide and isoalloxazine and flavin compounds in general. The bioactivity of the pigment situates it a worthy candidate as a biocontrol agent of other microbes. In light of this, various antimicrobial studies expounding on the bioactivity of pyocyanin-secretory strains of *P. aeruginosa* have been conducted.

Research exploits have traced microbial antagonistic activities of pyocyanin to be due to its redox activity in terms of univalent oxidoreductive ability as well as its ability to generate reactive oxygen species (ROS) and radicals of superoxide. This biomechanism is basically broad spectrum; however, extensive research has been embarked upon to enhance modes of delivery using nanotechnology and other important techniques, thus boosting its value for application [33]. In addition, the ROS generated by pyocyanin can be targeted against tumor cells, these cells are susceptible to pyocyanin-generated ROS as it basically interferes with some inherent intracellular eukaryotic cell functions like the activities of topoisomerase I and II.

An extensive industrial application of the electrochemical values of pyocyanin has also found applications in biosensor design and production. Pyocyanin is utilized as a redox compound for transporting electrons extracellularly between cells/test enzyme molecules and the material used as electrode. The application of such biosensor can be tailored for agriculture, environmental studies, and medical diagnostics.

In the textile industry, there is potential for the applications of phenazine pigments as textile colorants. This can be a way of reducing cost and increasing awareness on the sustainable use of natural products in industrial processes.

4. Contributions of *Pseudomonas aeruginosa* in the environment

A vast variety of synthetic chemicals have gained entrance into the ecosystem as a consequence of industrial activities, agricultural applications, and domestic usage which results to pollution in the environment. Furthermore, it is important to state that environmental pollution takes place when pollutants contaminate the surroundings, bringing about changes which adversely affect our normal lifestyles. This results in the incorporation of unwanted substances in the soil and water bodies due to anthropogenic activities. These pollutants are a result of oil spillage, herbicides, and pesticides. The use of herbicides and pesticides in soil ecosystem leads to the absorption of the nitrogen compounds into the soil. In response to this, nutrients might be available in the soil but not available to the plants.

4.1 Biodegradation of xenobiotics by various strains of *Pseudomonas aeruginosa*

Pesticides are organic compounds manufactured and used for the control of pests. They have great impact on human health and have active ingredient for the management and control of plant pests, insects and vermin [34]. An organochlorinated cyclodiene pesticide generally known as endosulfan (6,7,8,9,10,10-hexachloro-1,5,5a,6,9,9a-hexahydro-6,9-methano-2,4,3-benzodioxathiepine 3-oxide), currently in use around the globe for pest control in food and nonfood crops, has been classified under the category of persistent organic pollutant (POP). The solubility of endosulfan in water is significantly low, but it persists in the soil and water environment for more than 3 to 6 months. Endosulfan is pervasive and environmentally persistent; as a result the presence of endosulfan residues can be traced in surface water, groundwater, atmosphere, and water bodies. When these pesticides are released in the environment, they become pollutants, with ecological effects that require remediation. The consistent use and release of large quantities of pesticides either from accidental spills, direct application in agriculture, residues from cleaning of containers, faulty equipment, or inefficient methods used in the application of the products have led to huge environmental pollution and other attendant issues. Some of such issues include but not limited to changes in the nature of soils, groundwater, inland and coastal waters, and air [35, 36]. Biodegradation of toxic waste offers a promising strategy by which such chemicals may be detoxified [37] (**Figure 1**).

Soil microorganisms including *Pseudomonas* that are repeatedly exposed to pesticide may develop new capabilities to degrade such chemicals. Research studies have revealed that microbial degradation process to detoxify pesticide contaminations can be used effectively to overcome the pollution problems [38].

Pseudomonas aeruginosa play a vital role in the biodegradation and bioremediation of these toxic compounds found in soil and water by utilizing the pesticides as its carbon source and energy. Therefore *Pseudomonas aeruginosa* hold a lot of promises in the biodegradation of chemicals of environmental concerns into innocuous substances. In developed and developing countries, strains of *Pseudomonas*

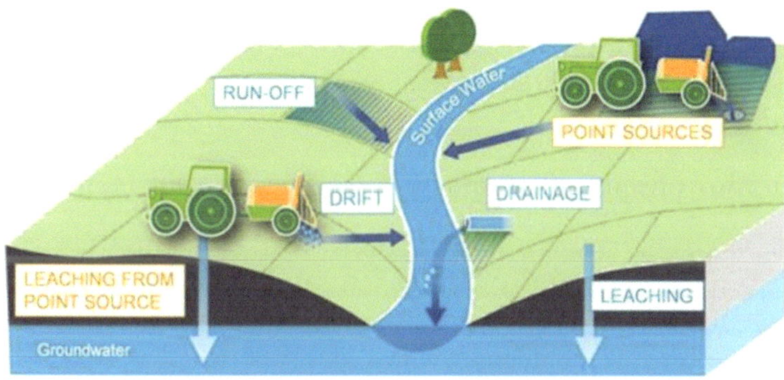

Figure 1.
Pesticide transport pathways to surface and groundwater in agricultural landscapes (www.agro.basf.com).

aeruginosa have been extensively employed in ex situ remediation. Herbicides and pesticides impacted soil which constitutes hazards if allowed to circulate in the food web and chain.

4.2 The use of *Pseudomonas aeruginosa* in the biodegradation and bioremediation of petroleum hydrocarbons

Crude oil spills into marine (offshore) bodies or soil (onshore) environments are very toxic and dangerous to the ecosystem and could be detrimental to the well-being of life forms, air, water, and soil processes and could as well increase the potential of fire hazards [39]. Onshore spill of crude oil affects living forms in the habitat, reduces agricultural productivity, and pollutes groundwater and sources of potable water and living biota in flowing water bodies, among others [40]. Eliminating or limiting these adverse effects from crude oil spillage situations implies total prevention of the spillage where possible and amending the soil via the procedure known as bioremediation [41]. Some known methods used in remediating crude oil-polluted soils include physical separation, chemical degradation, photodegradation, and bioremediation [42]. However, bioremediation is gaining preference because of its comparative effectiveness, relatively lower costs, and eco-friendliness when compared to other techniques. Conversely, methods other than bioremediation used for oil-polluted soil remediation have shown potentials of leaving secondary metabolites, which are secondary residuals left after the primary crude oil pollutant has been removed [43]. These by-products can even exhibit higher toxicity levels than the parent crude oil pollutant. Fortunately, bioremediation technique usage detoxifies contaminants in crude oil and effectively removes pollutant by destroying them instead of transferring them to other medium [44].

Researchers have used plant species for bioremediation, in a process known as phytoremediation, but the deploying microorganisms as biologically mediated remediation of crude oil-polluted soil are still linked to the effectiveness of phytoremediation systems [45]. This is as result of the fact that microorganisms are still required in the rhizosphere of plants for efficient soil remediation via phytoremediation [58]. This makes the use of microorganisms for the remediation of soil polluted by crude oil spills of increasing interest to researchers and stakeholders involved in crude oil-polluted soil amendment. A good example of bacterial strains of microbes used in reported works for effective repair of crude oil-polluted soil is *Pseudomonas aeruginosa* [46]. In an attempt to degrade petroleum hydrocarbon,

Pseudomonas will normally release biosurfactant which will reduce crude oil from high molecular weight to low molecular weight. This is to enable the bacteria to utilize the petroleum hydrocarbon [47].

4.3 Production of biosurfactants by *Pseudomonas aeruginosa*

Biosurfactants are a structurally diverse group of surface-active substances produced by microorganisms. The microorganisms that produce biosurfactants include *Pseudomonas*, *Bacillus*, *Micrococcus*, *Mycobacterium*, *Rhodococcus*, etc.

All biosurfactants are amphiphiles which consist of two parts: a polar (hydrophilic) moiety and nonpolar (hydrophobic) group [48].

The advantages of biosurfactants over chemically synthesized surfactants includes but not limited to; pH tolerance, less toxicity to the environment, biodegradability, better foaming properties, and them being able to be produced from agro-based industrial wastes [49].

Generally, biosurfactants have an ability to stabilize emulsions in various industrial applications [50] and are well-used in the food and pharmaceutical industries to achieve stability of emulsions. In addition, they have been applied in polluted water and soil during bioremediation in order to reduce interfacial tension, and it enhances the polar and nonpolar moieties to mix up.

Rhamnolipids are the major type of biosurfactant produced by *Pseudomonas aeruginosa* strain. Rhamnolipids are well-studied glycolipids secreted by *Pseudomonas aeruginosa* and have been found to have excellent surface activity [25] (**Figure 2**).

Incorporating rhamnolipids into remediation process enhances the solubility and elimination of these contaminants by improving oil biodegradations rates. Comparative study of biosurfactants for washing soil contaminated with crude oil was carried out where rhamnolipids showed a high degradable capacity; 80% of oil were degraded. Oil washing experiments by a combination of 10 g/l NaCl, 5.0 g/l n-butyl alcohol, and 2.0 g/l rhamnolipid provide very high oil extraction rates [52–54]. Even though rhamnolipids are the preferred enhancers for petroleum hydrocarbon soil pollutant degradation and have shown potentials to facilitate the bioremediation of soil contaminated by hydrocarbons, it has been suggested that their application must be evaluated carefully to reduce their exhibition on antimicrobial activity [55–57].

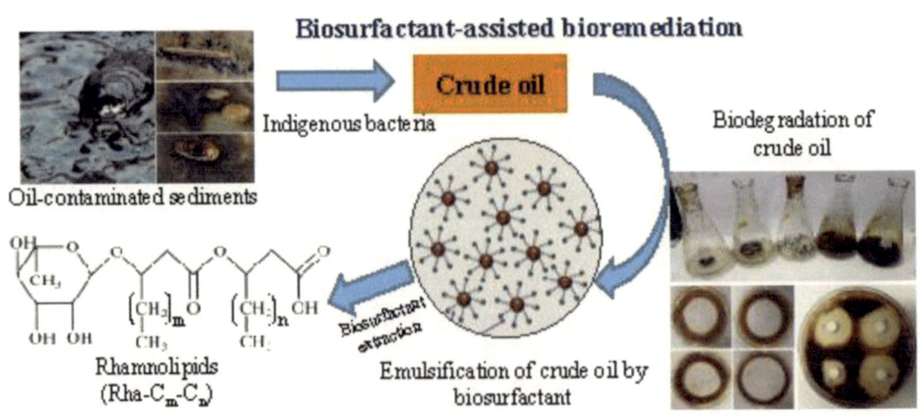

Figure 2.
Biosurfactant-assisted bioremediation of crude oil by indigenous bacteria. Isolated from Taean beach sediment [58].

5. Conclusion

Basically, strains of *Pseudomonas aeruginosa* have been widely implicated as clinical pathogens both in humans and in veterinary cases. In addition, they have been identified to be the causative agents of wound sepsis, septicemia, and nosocomial infections.

However, this chapter took a different dimension toward the beneficial roles of *Pseudomonas aeruginosa* strains in medicine, industries, and environment. The hopes of a clean environment through biodegradation of xenobiotics and bioremediation of hydrocarbon-impacted ecosystems are high with the use of *Pseudomonas aeruginosa*. In the industries, *Pseudomonas aeruginosa* holds a lot of practical promises toward production of intermediate products including metabolites such as rhamnolipids, vanillins, lipases, biopigments, etc.

Typically, in Nigeria and other developing countries, most of these materials from *Pseudomonas aeruginosa* are imported, and this does not encourage growth and development. Thus, local production of these intermediate products from *Pseudomonas aeruginosa* is adequate to save foreign reserves and promote development.

Author details

Orji Frank Anayo[1*], Onyemali Chidi Peter[1], Ukaegbu Gray Nneji[1], Ajunwa Obinna[3], Ezeanyanso Chika Scholastica[2] and Lawal Oluwabusola Mistura[1]

1 Department of Biotechnology, Federal Institute of Industrial Research, Lagos, Nigeria

2 Department of Chemical, Fiber and Environmental Technology, Federal Institute of Industrial Research, Lagos, Nigeria

3 Department of Microbiology, Modibbo Adama University of Technology, Yola, Adamawa, Nigeria

*Address all correspondence to: orjifa@yahoo.com

IntechOpen

© 2019 The Author(s). Licensee IntechOpen. This chapter is distributed under the terms of the Creative Commons Attribution License (http://creativecommons.org/licenses/by/3.0), which permits unrestricted use, distribution, and reproduction in any medium, provided the original work is properly cited.

References

[1] Palleroni N. The Pseudomonas story. Environmental Microbiology. 2010;**12**:1377-8310

[2] Butler M, Blaskovich M, Cooper M. Antibiotics in the clinical pipeline. The Journal of Antibiotics. 2013;**66**:571-591

[3] Bérdy J. Bioactive microbial metabolites. The Journal of Antibiotics. 2005;**58**:1-26

[4] Depoorter E, Bull M, Peeters C, Coenye T, Vandamme P, Mahenthiralingam E. Burkholderia: An update on taxonomy and biotechnological potential as antibiotic producers. Applied Microbiology and Biotechnology. 2016;**100**:5215-5229

[5] Stierle A, Stierle D. Bioactive secondary metabolites produced by the fungal endophytes of conifers. Natural Product Communications. 2015;**10**:1671-1682

[6] Thi Q, Tran V, Mai H, Le C, Hong M, Murphy B. Secondary metabolites from an Actinomycete from Vietnam's East Sea. Natural Product Communications. 2016;**11**:401-404

[7] Onbasli D, Aslim B. Determination of antibiotic activity and production of some metabolites by *Pseudomonas aeruginosa* B1 and B2 in sugar beet molasses. African Journal of Biotechnology. 2008;**7**:4814-4819

[8] Michel-Briand Y, Baysse C. The pyocins of *Pseudomonas aeruginosa*. Biochimie. 2002;**5**:499-510

[9] Cardozo V, Oliveira A, Nishio E, Perugini M, Andrade C, Silveira W. Antibacterial activity of extracellular compounds produced by a Pseudomonas strain against methicillin-resistant *Staphylococcus aureus* (MRSA) strain. Journal of Clinical Microbiology. 2013;**17**:12-24

[10] Chapalamadugu S, Chaudhry G. Microbiological and biotechnological aspects of metabolism of carbamates and organophosphates. Critical Reviews in Biotechnology. 1992;**12**:357-389

[11] Kerbauy G, Vivan A, Simões G, Simionato A, Pelisson M, Vespero E. Effect of metalloantibiotic produced by *Pseudomonas aeruginosa* on *Klebsiella pneumonia* Carbapenemase (KPC) – Producing *K. pneumoniae*. Current Pharmaceutical Biotechnology. 2016;**17**:389-397

[12] Simionato A, Navarro M, Jesus M, Barazetti A, Silva C, Simoes G. The effect of phenazine-1-carboxylic acid on mycelial growth of botrytis cinerea produced by *Pseudomonas aeruginosa* LV strain. Frontiers in Microbiology. 2017;**8**:1102

[13] Allouche N, Damak M, Ellouz R, Sayadi S. Use of whole cells of *Pseudomonas aeruginosa* for synthesis of the antioxidant hydroxytyrosol via conversion of tyrosol. Applied and Environmental Microbiology. 2004;**70**:2105

[14] Shimazaki M, Hasegawa J, Kan K, Nomura K, Nose Y, Kondo H, et al. Synthesis of captopril starting from an optically active β-hydroxy acid. Chemical & Pharmaceutical Bulletin. 1982;**30**:3139-3146

[15] Morahem A, Iraj N, Sayyed HZ, Fariborz M. Use of growing cells of *Pseudomonas aeruginosa* for synthesis of the natural vanillin via conversion of isoeugenol. Iranian Journal of Pharmaceutical Research (IJPR). 2011;**10**(4):749-757

[16] Cabrera-Valladares N, Richardson A, Olvera C, Treviño L, Déziel E, Lépine F, et al. Monorhamnolipids and 3-(3- hydroxyalkanoyloxy) alkanoic acids (HAAs) production

using *Escherichia coli* as a heterologous host. Applied Microbiology and Biotechnology. 2006;**73**:187

[17] Lang S, Wullbrandt D. Rhamnose lipids–biosynthesis, microbial production and application potential. Applied Microbiology and Biotechnology. 1999;**51**:22-32

[18] Ochsner U, Fiechter A, Reiser J. Isolation, characterization, and expression in *Escherichia coli* of the *Pseudomonas aeruginosa* rhlAB genes encoding a rhamnosyltransferase involved in rhamnolipid biosurfactant synthesis. The Journal of Biological Chemistry. 1994;**269**:19787-19795

[19] Ondetti M, Rubin B, Cushman D. Design of specific inhibitors of angiotensin-converting enzyme: New class of orally active antihypertensive agents. Science. 1977;**196**:441-444

[20] Soberón-Chávez G, Aguirre-Ramírez M, Sánchez R. The *Pseudomonas aeruginosa* RhlA enzyme is involved in rhamnolipid and polyhydroxyalkanoate production. Journal of Industrial Microbiology & Biotechnology. 2005;**32**(11-12):675-677

[21] Pornsunthorntawee O, Wongpanit P, Chavadej S, Abe M, Rujiravanit R. Structural and physicochemical characterization of crude biosurfactant produced by *Pseudomonas aeruginosa* SP4 isolated from petroleum—Contaminated soil. Bioresource Technology. 2008;**99**:1589-1595

[22] Van H, Vanzeveren E. Rhamnolipids in bakery products; 2004

[23] Nitschke M, Costa S. Biosurfactants in food industry. Trends in Food Science and Technology. 2007;**18**:252-259

[24] Muller M, Kugler J, Henkel M, Gerlitzki M, Hormann B. Rhamnolipids-next generation surfactants. Journal of Biotechnology. 2012;**162**:366-380

[25] Irfan-Maqsood M, Seddiq-Shams M. Rhamnolipids-well-characterized glycolipids with potential broad applicability as biosurfactants. Industrial Biotechnology. 2014;**10**: 285-291

[26] Vatsa P, Sanchez L, Clement C, Baillieul F, Dorey S. Rhamnolipid biosurfactants as new players in animal and plant defense against microbes. International Journal of Molecular Sciences. 2010;**11**:5095-5108

[27] Vadlamani S, Parcha S. Studies on industrially important alkaline protease production from locally isolated superior microbial strain from soil microorganisms. International Journal of Biotechnology Applications. 2011;**3**:102-105

[28] Samanta A, Pal P, Mandal A, Sinha C, Lalee A, Das M, et al. Estimation of biosurfactant activity of an alkaline protease producing bacteria isolated from municipal solid waste. Central European Journal of Experimental Biology. 2012;**1**:26-35

[29] Lee J, Boyapati G, Song K, Rhee S, Kim C. Cloning and sequence analysis of the estA gene encoding enzyme for producing(R)-beta-acetylmercaptoisobutyric acid from *Pseudomonas aeruginosa* 1001. Journal of Bioscience and Bioengineering. 2000;**90**:684-687

[30] Sharma R, Chisti Y, Banergee U. Production, purification, characterization, and applications of lipases. Biotechnology Advances. 2001;**19**:627-662

[31] Priyaja A. Pyocyanin (5-methyl-1-hydroxyphenazine) produced by *Pseudomonas aeruginosa* as antagonist to vibrios in aquaculture: Overexpression, downstream process and toxicity

(Ph.D thesis). India: Cochin University of Science and Technology; 2013

[32] El-Fouly MZ, Sharaf AM, Shanin AAM, El-Baily HA, Omara AMA. Biosynthesis of pyocyanin pigment by *Pseudomonas aeruginosa*. Journal of Radiation Research and Applied Science. 2015;**8**:36-48

[33] De Vleesschauwer D, Cornelis P, Hofte M. Redox-active Pyocyanin secreted by *Pseudomonas aeruginosa* 7NSk2 triggers systemic resistance to *Magnaporthe grisea* but enhances *Rhizoctonia solani* susceptibility in rice. MPMI. 2006;**19**(12):1406-1419

[34] Aghilinejad M, Farshad A, Naghavi M, Haghani H. Assessment of the relationship between pesticide and their effects on farmer health in various state. Iran Occupational Health. 2006;**3**(1):81-85

[35] Bhandhade B, Satnaik S, Kanetar P. Bioremediation of an industrial effluent containing monocrotophos. Microbiologica. 2002;**45**:346-349

[36] Ortiz-Hernández M, Sánchez-Salinas E. Biodegradation of the organophosphate pesticide tetrachlorvinphos by bacteria isolated from agricultural soils in Mexico. Revista Internacional de Contaminación Ambiental. 2010;**26**:27-38

[37] Manisha D, Shyamapada M, Nishith K. Plasmid-mediated dimethoate degradation by *Bacillus licheniformis* isolated from a fresh water fish Labeo rohita. Journal of Biomedicine & Biotechnology. 2005;**3**:280-286

[38] Kumar M, Philip L. Bioremediation of endosulfan contaminated soil and water: Optimization of operating conditions in laboratory scale reactors. Journal of Hazardous Materials. 2006;**136**:354-364

[39] Abioye O. Biological remediation of hydrocarbon and heavy metals contaminated soil. In: Pascucci S, editor. Soil Contamination. IntechOpen; 2011. pp. 127-142

[40] Perelo L. Review: in situ and bioremediation of organic pollutants in aquatic sediments. Journal of Hazardous Materials. 2010;**177**:81-89

[41] Chhatre S, Purohit H, Shanker R, Chakrabarti T, Khanna P. Bacterial consortia for crude oil spill remediation. Science and Technology. 1996;**34**:187-193

[42] Liu Y, Li C, Huang L, He Y, Zhao T, Han B, et al. Combination of a crude oil-degrading bacterial consortium under the guidance of strain tolerance and a pilot-scale degradation test. Chinese Journal of Chemical Engineering. 2017;**25**:1838-1846

[43] Minoui S, Minai-Tehrani D, Shahriari M. Phytoremediation of crude oil-contaminated soil by Medicago sativa (alfalfa) and the effect of oil on its growth. In: Öztürk M, Ashraf M, Aksoy A, Ahmad MSA, editors. Phytoremediation for Green Energy. Dordrecht: Springer Science + Business Media; 2015. pp. 123-129

[44] Okoh A, Trejo-Hernandez M. Remediation of petroleum polluted systems: Exploiting the bioremediation strategies. African Journal of Biotechnology. 2006;**5**:2520-2525

[45] Desai J, Banat I. Microbial production of surfactants and their commercial potential. Microbiology and Molecular Biology Reviews. 1997;**61**:47-64

[46] Akhundova E, Atakishiyeva Y. Interaction between plants and biosurfactant producing microorganisms in petroleum contaminated Absheron soils. In: Öztürk M, Ashraf M, Aksoy A, Ahmad MSA, editors. Phytoremediation for Green Energy. Dordrecht: Springer Science+Business Media; 2015. pp. 115-122

[47] Zhang X, Xu D, Zhu C, Lundaa T, Scherr K. Isolation and identification of biosurfactant producing and crude oil degrading *Pseudomonas aeruginosa* strains. Chemical Engineering Journal. 2012;**209**:138-146

[48] Pacwa-Plociniczale M, Plaza GA, Piotrowska-Seget Z, Cameotra SS. Environmental applications of biosurfactants : Recent advances. International Journal of Molecular Sciences. 2011;**12**(1):633-654

[49] Sifour M, Al-Jiwali MA, Aziz GM. Emulsification properties of biosurfactants produced from *Pseudomonas aeruginosa* RB 28. Pakistan Journal of Biological Sciences. 2007;**10**(8):1331-1335

[50] Obayori OS, Ilori MO, Adebusoye SA, Oyetibo GO, Omotayo AE, Amund OO. Degradation of hydrocarbons and biosurfactant production by *Pseudomonas* sp. strain LP1. World Journal of Microbiology and Biotechnology. 2009;**25**:1615-1623

[51] Dong L, Hanbyu L, Bong-Oh K, Jong S, Un Hyuk Y, Beom S, et al. Biosurfactant-assisted bioremediation of crude oil by indigenous bacteria isolated from Taean beach sediment. Environmental Pollution. 2018;**241**:254-264

[52] Beal R, Betts W. Role of rhamnolipid biosurfactants in the uptake and mineralization of hexadecane in *Pseudomonas aeruginosa*. Journal of Applied Microbiology. 2000;**89**:158-168

[53] Cameotra S, Singh P. Bioremediation of oil sludge using crude biosurfactants. International Biodeterioration and Biodegradation. 2008;**62**:274-280

[54] Zheng C, Wang M, Wang Y, Huang Z. Optimization of biosurfactant-mediated oil extraction from oil sludge. Bioresource Technology. 2012;**110**:338-342

[55] Kaczorek E. Effect of external addition of rhamnolipids biosurfactant on the modification of gram positive and gram negative bacteria cell surfaces during biodegradation of hydrocarbon fuel contamination. Polish Journal of Environmental Studies. 2012;**21**:901-909

[56] Rahman K, Rahman T, Kourkoutas Y, Petsas I, Marchant R, Banat I. Enhanced bioremediation of n-alkane in petroleum sludge using bacterial consortium amended with rhamnolipid and micronutrients. Bioresource Technology. 2003;**90**:159-168

[57] Zezzi-Gomes M, Nitschke M. Evaluation of rhamnolipid and surfactin to reduce the adhesion and remove biofilms of individual and mixed cultures of food pathogenic bacteria. Food Control. 2012;**25**:441-447

[58] Mohsenzadeh F, Rad A. Bioremediation of petroleum polluted soils using *Amaranthus retroflexus* L. and its rhizospheral fungi. In: Öztürk M, Ashraf M, Aksoy A, Ahmad MSA, editors. Phytoremediation for GreenEnergy. Dordrecht: Springer Science+Business Media; 2015. pp. 131-139